T0135711

Machine Learning in Advanced Driver-Assistance Systems

Contributions to Pedestrian Detection and Adversarial Modeling

von der Fakultät für Elektrotechnik, Informationstechnik und Medientechnik der
Bergischen Universität Wuppertal
genehmigte

Dissertation

zur Erlangung des akademischen Grades
eines Doktors der Ingenieurwissenschaften

von
M.Sc. Farzin Ghorban Rajabizadeh
aus
Wuppertal

Wuppertal 2019

Tag der mündlichen Prüfung: 18. Januar 2019
Hauptreferent: Prof. Dr.-Ing. Anton Kummert
Korreferent: Prof. Dr.-Ing. Reinhard Möller

Studien zur Mustererkennung

herausgegeben von:

Prof. Dr.-Ing. Heinrich Niemann
Prof. Dr.-Ing. Elmar Nöth

Bibliografische Information der Deutschen Nationalbibliothek

Die Deutsche Nationalbibliothek verzeichnet diese Publikation in der
Deutschen Nationalbibliografie; detaillierte bibliografische Daten sind
im Internet über http://dnb.d-nb.de abrufbar.

ISBN 978-3-8325-4874-2
ISSN 1617-0695

Logos Verlag Berlin GmbH
Comeniushof
Gubener Str. 47
10243 Berlin
Tel.: +49 030 42 85 10 90
Fax: +49 030 42 85 10 92
INTERNET: http://www.logos-verlag.de

Acknowledgements

First and foremost I wish to express my sincerest gratitude and thanks to my advisor Prof. Anton Kummert, who accompanied me during my thesis journey and was welcoming at all times. He has been a great mentor and this work would not have been possible without his excellent scientific views as well as human qualities. I am honored to have been his student. I also want to express my gratitude to Christian Nunn, who gave me the opportunity to work on many wonderful, interesting, and challenging topics during my thesis. Most importantly, he gave his commitment and support whenever it was needed. The last years at Delphi were very memorable and significant. I am fortunate to have been a part of such a fantastic team. The atmosphere was outstanding in all aspects. I have learned a lot from many wonderful people. This included exhaustively exploring different methodologies, analyzing results, and properly communicating the findings. I deeply thank Alessandro Colombo, my research mentor during my first year at Delphi, for the great and insightful discussions we have had. I wish to extend my thanks to all my research mentors at Delphi over the years for their consistent support and encouragements: Mirko Meuter, Javier Marin, Yu Su, Dennis Müller, Lutz Roese-Koerner. My office mate: Maryam Foroughi, Daniel Schugk, Farnoush Zohourian, Stephanie Lessmann, Jens Westerhoff, and Narges Milani. I am grateful. Thank you for your support, personal, and professional understanding. Finally, I wish to acknowledge the support I have received from my parents, brothers, and wife. Thank you all so much for supporting me and creating a peaceful environment. I am ever grateful. Last words go to Nesreen, my wife. Thank you for your patience, perseverance, and standing by my side.

Contents

List of Tables

List of Figures

List of Algorithms

List of Abbreviations

ACF	Aggregated channel features.
ACNet	Aggregated channels network.
AdaBoost	Adaptive boosting.
Adam	Adaptive moment estimation.
ADAS	Advanced driver-assistance systems.
AP	Average precision.
Bagging	Bootstrap aggregating.
BF	Boosted forest.
CBF	Cascaded boosted forest.
CNN	Convolutional neural network.
CPU	Central processing unit.
CSS	Color self similarity.
D	Discriminator.
FPPI	False positives per image.
FPS	Frames per second.
G	Generator.
GANs	Generative adversarial networks.
GPU	Graphics processing unit.
HOG	Histogram of oriented gradients.
ICBF	Insatiate cascaded boosted forest.
ICF	Integral channel features.

IoU   Intersection over union.

LUV   Color space based on luminance, L, and chromaticity coordi-
                   nates, U and V.

MCGANs   Multichannel generative adversarial networks.

MPC   McCulloch-Pitts cell.

MR   Log-average miss rate.

NMS   Non-maximum suppression.

pp   Percent points.

ReLU   Rectified linear unit.

ResNet   Residual network.

RGB   Color space based on red, green, and blue colors.

ROC   Receiver operating characteristic.

RoI   Region of interest.

RPN   Region proposal network.

SI   Sampling interval.

SVM   Support vector machine.

YUV   Color space based on luma, Y, and chromaticity coordinates,
                   U and V.

Chapter 1

Introduction

Nowadays computers are ubiquitous. They contribute to making our lives more convenient and secure. They have the potential ability to save human lives, which is well demonstrated in their deployment in modern vehicles. In the context of advanced driver-assistance systems (ADAS), vehicles are equipped with multiple sensors including lidar, radar, and camera all of which record the vehicle's environment in addition to intelligent algorithms for analyzing and understanding the recorded data. For understanding the vehicle's environment, ADAS unite multiple modules such as forward collision detection [135], obstacle detection [182], lane guidance [179], traffic sign recognition [114], and pedestrian detection [185]. Statistics show that over 90 percent of road accidents occur due to human errors[1]. A vehicle's ADAS can, in advance, alert the driver of hazardous conditions or actively intervene in such situations to reduce the human error and potentially reduce road accidents.

This study contributes to the research in modern ADAS on different aspects. The two main contributions comprise both pedestrian detection, that is recognizing and localizing pedestrians in images, and synthetic traffic sign generation. In chapters 2 and 3, we outline relevant research on object detection then discuss methodologies and data that are used to rank our approaches and compare them to the state of the art.

Methods deployed in ADAS must be accurate and computationally efficient in order to run fast. Ideally, they are required to execute in real time on embedded platforms. In chapter 4 and [66, 62], we introduce a novel approach for pedestrian detection that is specially designed for low-consumption hardware. Concretely, we identify the proposal evaluation phase as the computational bottleneck of two-stage cascades that involve a

---

[1]https://www.dekra-roadsafety.com/media/dekra-verkehrssicherheitsreport-2016-de.pdf

convolutional neural network (CNN) as the second component. As for the first component, we employ a cascaded boosted forest (CBF) detector. In order to economize on the computational cost of the arrangement, we share the feature pyramid that the CBF detector constructs and forward only features that belong to the promising locations in the image to the CNN classifier. In this manner, the expensive feature computation is done once and features are reused by the CNN. For evaluating the features, we design a small-sized CNN that can rapidly process the small proposal dimensions and has a sufficient depth to achieve an accurate classification quality. The CNN is trained from scratch. In various evaluations its optimal operational point, training routine, and location in the pipeline are determined. We demonstrate that our approach can achieve a high performance while running in real time with 30 frames per second without being parallelized and without the use of a GPU. Furthermore, we introduce multiple versions of our approach. The results concluded that our three-stage cascade ranks as the fourth best-performing method reported on one of the challenging pedestrian datasets that are available online.

The other challenge we face with ADAS would be the issue of training efficient detection methods which requires human effort. This would be an extensive manual annotation for preparing training data. In chapter 5 and [65], we introduce a novel approach and insights to make CBF detectors a more data efficient. We decompose a detector into its fundamental parts in order to obtain a better understanding of how the different components contribute to the detection quality. A crucial insight from our evaluations is that the underlying AdaBoost algorithm in CBF frameworks not only copes with highly imbalanced numbers of positive (pedestrians) and negative (backgrounds) training samples but it also benefits from a relatively high number of negative samples. This insight is relevant for many multi-scale object detection tasks since the number of available positive samples in datasets usually is a fraction of the number of the negative samples. In order to exploit the asymmetry in the datasets, however, it is essential to optimize the training routine, especially the sample selection and gathering process. We propose an approach for gathering a sufficient number of high-quality samples without the need for any data augmentation technique. We demonstrate that our approach effectively prevents overfitting and, therefore, allows increasing the model capacity without incurring the risk of performance reduction or poor generalization. Our approach is orthogonal to known researches and can, therefore, be employed in existing CBF detection methods without decelerating the detector. We demonstrate comparisons to the state of the art where we rank as second-best among CBF detectors on two challenging pedestrian datasets. This is achieved while using a relatively small number of simple aggregated channel features, which allows our detector to run multiple times faster than competitors.

Acquiring labeled training data is costly and time-consuming, particularly in the case of traffic sign recognition, since countries do not use unified traffic signs plus different traffic signs do not occur equally often. Due to these difficulties, it requires many hours of acquisition and preparation to obtain a large number of well-balanced and labeled training samples. In chapter 6 and [63, 64], we investigate the use of synthetic data and the involvement of advanced learning approaches with the aspiration to reduce the human efforts behind the data preparation and to make the training of recognition models more data efficient. For these purposes, we employ the approach of generative adversarial networks (GANs). Our study comprises two contributions. Primarily, we algorithmically and architecturally adapt the adversarial modeling framework to the image data provided in ADAS, the so-called red-clear-clear-clear images. We demonstrate that our framework can process multiple channels that have different resolutions and textures, and generate real-looking red-clear-clear-clear traffic sign samples. Our framework allows adaptation of known approaches that we use to enable the generator to create specific samples and even to change incisive attributes of the samples. We also demonstrate that a variation of our framework can transfer visual properties. Secondarily, we study and discuss relevant researches that successfully employ synthetic data for training traffic sign recognition models. Based on the studies and detailed analyses and evaluations of our framework, we discuss future research directions and conclude that GANs can contribute in multiple ways to the training of traffic sign classifiers.

Chapter 7 concludes this work with a summary, discussion, and perspectives for future research.

Chapter 2

Machine Learning for Object Detection

2.1 Introduction

Object detection is one of the most important disciplines in image understanding. Detection methods are required for localizing an object of interest within an image. With advancements in computer vision, numerous detection frameworks have been developed. This chapter provides a brief overview of recent methods for object detection with a special focus on pedestrian detection under real-world conditions [23, 11].

Systematic overview. A majority of multiscale detection methods discussed in this study can be decomposed into fundamental subprocesses as shown in figure 2.1. Some methods may further include pre- or postprocessing steps, employ the subprocesses in a different order, or omit some subprocesses. In the following, we briefly describe the major tasks of each subprocess and review them in greater detail in the coming sections:

- Proposal generation: Classification methods predict the class membership of a given patch, which usually has a predefined dimension. Proposal generation methods define the search space for a classifier. In other words, proposal generation methods present patches from different locations and scales in an image to the classifier and reformulate the localization task as, for example, an iterative classification task.

- Feature creation: Classification methods use feature creation functions to map the image/patch into a feature space where the most relevant characteristics of the image/patch are emphasized. These character-

5

istics have various levels of complexity and can include shapes, colors, edges, abstract information, etc.

- Classification: Both processes mentioned above manage the input stream to the classifier. In the case of pedestrian detection, the classifier functions in a binary manner and discriminates the input patches between pedestrians (positives) and backgrounds (negatives). The locations of the classified pedestrians are marked, for example, using bounding boxes that surround the pedestrians.

- Bounding box clustering: The number of detections (bounding boxes) is usually weakly correlated with the number of objects in an image. To understand the contents of an image, a method is often employed to cluster neighboring detections and remove redundant boxes.

- Bounding box regression: Similar to classification methods, a regression method receives an input patch but predicts its correct bounding box location.

The remainder of this chapter is organized as follows. In section 2.2, we outline some of the most important machine learning techniques used for

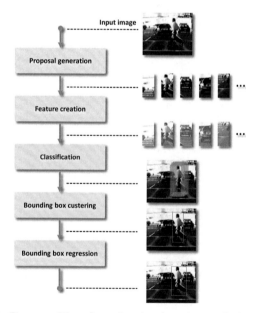

Figure 2.1: Decomposition of a pedestrian detection method in its fundamental subprocesses.

object detection and regression in images. In section 2.3, we review relevant state-of-the-art mechanisms for proposal generation. Section 2.4 describes feature representations with a special focus on those used in our study and finally, in section 2.5, post-processing algorithms used for bounding box clustering are discussed.

2.2 Machine learning

For solving some real-world problems, it is required to find a complex function that maps an input x into some desired output y. Machine learning approaches train a model that approximates such a function, as closely as possible, without being explicitly programmed or guided by rules but only implicitly through a set of samples [147]. This set is referred to as *training set* and is composed of N corresponding pairs $\mathcal{X} = \{x^{(1)}, \ldots, x^{(N)}\}$ and $\mathcal{Y} = \{y^{(1)}, \ldots, y^{(N)}\}$. Here, $x^{(i)}$ may represent an image patch, i.e., $x^{(i)} \in \mathbb{R}^{H^p \times W^p \times C^p}$, where H^p, W^p, and C^p refer to the height, width, and depth of the patch, respectively. The task is termed *classification* if $y^{(i)} \in \mathbb{N}^n$ with $n \geq 1$ (for $n > 1$, $y^{(i)}$ is usually one-hot encoded) and if $y^{(i)} \in \mathbb{R}^n$, the task is termed *regression*. The capability of the trained model to *generalize*, i.e., the ability to perform accurately on a set of new, unseen samples/tasks is an important property of these approaches and is sought to be maximized. The set of the new, unseen samples is referred to as the *test set*.

Usually, one distinguishes between three types of learning approaches [25]:

- Supervised learning: The training set comprises both $\mathcal{X}$ and the corresponding desired outputs $\mathcal{Y}$. During training, $\mathcal{X}$ and $\mathcal{Y}$ are presented to the model and the model parameters are adapted according to the distance between the produced and the desired outputs.

- Unsupervised learning: $\mathcal{X}$ is available but $\mathcal{Y}$ is not. $\mathcal{X}$ can be used, for example, to discover groups of similar samples within the training data, this is known as *clustering*.

- Reinforcement learning: The exact output of the function to be learned is unknown, and training relies on parameter adjustments based on two concepts: reward and penalty. In other words, if the model does not perform well enough, it is penalized and its parameters are adapted accordingly. Otherwise, it is rewarded, i.e., reinforcement occurs. The difference between reinforcement learning and supervised learning is that in reinforcement learning, optimal outputs must be discovered by a process of trial and error. Reinforcement learning is employed in dynamic environments and applications include, for example, learning to drive a vehicle or playing a game against an opponent.

2.2.1   Ensemble learning

Ensemble methods arrange multiple models to constitute a single model that demonstrates a better prediction performance than any of the constituent models. It is based on the idea that training several simple models and combining them into a complex model is easier than training a single complex model. In the following, we outline some of the most popular representatives of the two paradigms of ensemble methods [213]:

- Parallel modeling: Constituent models are generated in parallel and trained independently. Two approaches that are often used in parallel modeling are *Bagging* and *random forest*, which we describe in detail in sections 2.2.1.2 and 2.2.1.3, respectively.

- Sequential modeling: Constituent models are generated sequentially and each individual model trains while considering the performance of its precursors. In section 2.2.1.4, we describe the *boosting* approach which is often used in sequential modeling.

There are several views on both paradigms; however, one can generally say that both succeed by increasing the diversity among the constituent models [98]. The individual models are generated using either resampling (portioning new training sets), attributes resampling, or reweighting (reweighting training samples).

Most ensemble models conceptually accept any type of prediction models. However, since ensemble models typically require more computation than a single model, they are often applied in the literature using fast constituent models such as decision trees. Therefore, we review the foundations of decision tree models in the following section.

2.2.1.1   Decision tree

A *decision tree*, used for classification and regression problems [17], denotes a set of nodes that are connected through edges in a particular way. An example of such a structure is shown in figure 2.2. Each node has either one incoming edge, multiple outgoing edges, or both. The entry point to the

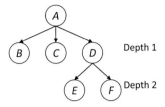

Figure 2.2: Illustration of a decision tree.

tree is the node that does not have any incoming edges, which is referred to as the *root node*. Terminal nodes are those without outgoing edges and are referred to as *leaf nodes*. The remaining nodes are known as *decision nodes* or *splits*. The *depth* of the tree is defined as the number of nodes from the root to the furthest node. The tree, shown in figure 2.2, has a depth of 2, {E, F} are the furthest nodes with two nodes between them and the root node, namely A and D. Trees with depth 1 are often called *decision stumps*. A decision node is referred to as *n*-ary if it has *n* outgoing edges. Any *n*-ary decision node can be represented as a tree consisting solely of 2-ary nodes — such types of trees are referred to as *binary trees*. A *univariate* tree applies axis-parallel splits. In other words, in each node, one feature is compared against one threshold and the outcome decides the path in which the sample travels downward to a terminal node. On the contrary, a *multivariate* tree is not restricted to splitting samples along the axis of the features.

Decision trees are usually built using a heuristic called *recursive partitioning* (also known as *divide and conquer*). This approach repeatedly splits the data into subsets at each decision node and grows the tree until: a) the data in the subsets (leaves) is sufficiently homogeneous, b) there is no feature left permitting further discrimination among the samples, or c) a predefined stopping criterion has been satisfied. There are numerous methods for finding features that best divide the data into subsets [116, 121]. The decision tree, as any learned hypothesis, tends to overfit the training data. There are two common strategies that most tree construction algorithms implement to avoid overfitting [92]. The first approach is *early stopping* (or *pre-pruning*) before the model perfectly fits the training data. This can be done by, for example, restricting the maximum depth of the tree or the minimum permissible error (*purity* of the leaves) that forbids further splitting when the error in the subset falls below a predefined value. An alternative to this approach is known as *post-pruning*. In post-pruning, the tree is first fully constructed and then leaves and nodes that are not efficient are pruned and the tree is reduced to an appropriate depth [46].

2.2.1.2   Bagging

Bootstrap. Bootstrap [44] is a statistical method that relies on *random sampling*. It assigns measures of accuracy (e.g., bias and variance) to sample estimates. Therefore, several (non-disjoint) sets are obtained by randomly drawing samples with replacement from a single training set $\mathcal{D}$. All instances are selected with a probability of $1/N$. The idea is to obtain B new sets $\mathcal{D}_{b \leq B} \subseteq \mathcal{D}$, each containing $n \leq N$ samples from $\mathcal{D}$. On each set $\mathcal{D}_b$, some statistic Ψ_b can be evaluated. An estimate of Ψ is given as $\tilde{\Psi} = \frac{1}{B} \sum_b \Psi_b$. Let $\hat{\Psi}$ denote Ψ evaluated on the base set $\mathcal{D}$. For any finite N and B, $\tilde{\Psi} \neq \hat{\Psi}$, where the difference is termed *bias*, which gives an idea of the expected deviation from the true value of Ψ. As a final result for Ψ we can quote $\Psi = \tilde{\Psi} \mp \sigma_{\tilde{\Psi}}$, where $\sigma_{\tilde{\Psi}}$ refers to the standard deviation, which is given as $\sigma_{\tilde{\Psi}}^2 = \frac{1}{B} \sum_b (\Psi_b - \tilde{\Psi})^2$.

According to Breiman [16], a learning method is termed *unstable* if a small change in its training data results in a large change in its prediction. In other words, these methods tend to suffer from random fluctuations in training data. Therefore, such models would have a large variance in their predictions if they are trained on different portions of the training data. In [14], Breiman proposed an ensemble learning method named *bootstrap aggregating* (known as Bagging) for such models. Bagging creates an ensemble of B predictors, where each is trained on one bootstrap set $\mathcal{D}_b$.

Algorithm 2.1 summarizes the Bagging procedure. Bagging accepts any type of predictors. However, Breiman demonstrated that the underlying models should be unstable (such as decision trees or neural networks) otherwise the procedure may degrade performance. Bagging is widely used in combination with decision tree models and as the decision tree itself, it can be used for regression and classification. For regression tasks, the responses of the models are averaged and for classification, the majority vote is used.

Algorithm 2.1 *Bagging*

Input:
- $\mathcal{D} = \{(x^{(i)}, y^{(i)})\}_{i \leq N}$
- B number of predictors $\varphi_{b \leq B}$
- n with $n \leq N$ number of samples to be used for each predictor

Output:
- $\{\varphi_1, \ldots, \varphi_B\}$ ensemble of B predictors

1: **for** $b = 1, \ldots, B$ **do**:
2:     Create $\mathcal{D}_b$ by randomly drawing n samples from $\mathcal{D}$ with replacement
3:     Create φ_b using samples from $\mathcal{D}_b$
4: **end for**

2.2.1.3   Random forest

Random subspace. In [75], a method was proposed for constructing an ensemble of tree classifiers that are trained using all training samples with different portions of available features, also known as *random subspace* and *attribute Bagging*. The basic idea behind the method is the use of stochastic perturbation and averaging to avoid overfitting [90, 58].

In [15], Breiman incorporated this idea into the Bagging algorithm. While Bagging does not consider the type of predictors created or the manner in which they are created, random forest is particularly designed to decrease the correlation in the created tree models [178, 43, 15]. This is achieved by considering only a subset of the available features for each split. After constructing the ensemble, votes are considered in the same manner as in the Bagging procedure. An advantage of random forest is that it is relatively insensitive to the size of the subset and because finding the best split is usually the costliest step in tree construction, random forest considerably reduces the construction time.

2.2.1.4   Boosting

Boosting denotes a class of machine learning methods capable of constructing a *strong learner* with arbitrarily high accuracy. This is achieved by linearly combining moderately accurate *weak learners* that perform slightly better than random guessing, i.e., with an associated error, $\epsilon < 0.5$. Schapire et al. [159] proposed a boosting procedure that uses a resampling mechanism and trains two additional classifiers on filtered versions of the data. For this procedure, the initial model, h_1, learns from N samples. h_2 uses a new set of the same length, where $N/2$ of the N samples are misclassified by h_1. Finally, h_3 learns on N samples for which h_1 and h_2 disagree. The three classifiers constitute a boosted classifier H, which decides based on a majority vote from all $h_{i \leq 3}$. It was shown in [159] that H outperforms each of the weak classifiers. This procedure can be run recursively, that is, each h_i can be replaced using a boosted classifier H to achieve enhanced performance [52].

AdaBoost. Freund and Schapire [57] proposed an *adaptive boosting* algorithm (known as AdaBoost), which demonstrates efficient performance by iteratively reweighting training samples. The idea is that initially there is nothing known about the training data. In other words, at iteration $t = 1$, a weight $w_{i \leq N}^{t=1} = 1/N$ is assigned to each sample, treating all samples equally. The algorithm presupposes that at each iteration, a weak learner with $\epsilon_t \leq 1/2 - \gamma_t$ where $\gamma_t > 0$ can be provided, i.e., there exists a weak

---

Algorithm 2.2 *AdaBoost for binary classification*

Input:
- T number of iterations
- $\mathcal{D} = \{(x^{(i)}, y^{(i)})\}_{i \leq N}$, with $y^{(i)} \in \{-1, 1\}$

Output:
- $H(x) = \sum_t \alpha_t h_t(x)$

Initialize:
- $w_i^1 = 1/N$

1: **for** $t = 1, \ldots, T$ **do:**
2:     Train weak learner $h_t : \mathcal{X} \to \{-1, 1\}$ using w^t
3:     Calculate error of h_t : $\epsilon_t = \sum_i^N w_i^t \mathbb{1}[h_t(x^{(i)}) \neq y^{(i)}]$
4:     Set $\alpha_t = \frac{1}{2}ln(\frac{1-\epsilon_t}{\epsilon_t})$
5:     Update weights $w_i^{t+1} = w_i^t e^{-\alpha_t y^{(i)} h_t(x^{(i)})}/Z_t$, where $Z_t = \sum_i^N w_i^{t+1}$
6: **end for**

---

classifier that is an *edge* (γ_t) better than random guessing. In the context of boosting, AdaBoost requires each subsequent classifier to focus on the wrong classified samples from its precursors. AdaBoost achieves this by adjusting the weights of the new iteration such that misclassified samples gain weight and correct classified samples lose weight. Each weak learner is then required to achieve a weighted error

$$\epsilon_t = \sum_i w_i^t \mathbb{1}[h_t(x^{(i)}) \neq y^{(i)}] \leq 1/2 - \gamma_t < 1/2, \qquad (2.1)$$

where $\mathbb{1}[x]$ is the indicator function that is equal to 1 for $x = true$ and 0 for $x = false$. Algorithm 2.2 summarizes a variant of AdaBoost, which is widely used for binary classification, known as *discrete AdaBoost* [55]. The algorithm is adaptive in the sense that it adapts, through α_t, to the error rates of the individual weak learner and therefore does not need to know γ ab initio.

AdaBoost guarantees that when weak learners are provided in accordance to the above definition and there exists a sufficient number of training samples, the error on the training set drops exponentially with [57]

$$\sum_i w_i^1 \mathbb{1}[y^{(i)} \neq H(x^{(i)})] \leq \prod_t Z_t = \prod_t 2\sqrt{\epsilon_t(1 - \epsilon_t)} \leq e^{-2\sum_t \gamma_t^2}. \qquad (2.2)$$

However, in general, a model that is too accurate on the training set might not be accurate on independent samples — it might overfit the training data. Restricting the number of iterations, T, and having simple weak learners usually helps avoid overfitting of the training data [56]. AdaBoost sometimes empirically demonstrates overfitting of the training data when the final model becomes overly complex, and in other experiments, it demonstrates that overfitting can but does not have to occur [160]. In the latter case, it is reported that the test error did not increase with increasing T, and the test error even dropped further after the training error reached zero. This observation has resulted in a new perspective on AdaBoost [162, 55]. The reason for such behavior has been explained by analyzing the margin,

$$y_i \frac{\sum_t \alpha_t h_t(x^{(i)})}{\sum_t \alpha_t}, \tag{2.3}$$

which translates to the confidence of the model and may continue to increase even after H accurately fits the training data. AdaBoost has been discussed, viewed, and adapted for different purposes [161]. Freud et al. [57] proposed variants of AdaBoost for multiclass and regression problems. In [163] authors extended AdaBoost to handle weak learners that output confidence-rated predictions, where each $h_t(x) \in \mathbb{R}$ whose sign is the predicted label and whose magnitude $|h_t(x)|$ provides a measure of confidence. This variant is called *real AdaBoost*. The simplicity of AdaBoost and its ability to cope with any weak learner makes it attractive, and therefore there exists a variety of methods based on the idea of AdaBoost [97, 52].

2.2.1.5 Cascading

Cascading is usually employed in sequential ensemble modeling and aims to improve the efficiency of such structures. It is essentially a simple-to-complex strategy that performs heavy computation only on promising locations. This is because, in most detection tasks, an overwhelming majority of locations that must be evaluated are usually negatives. Excluding those locations from complex parts of a framework results in faster evaluation time. The term cascading was coined based on the groundbreaking work of Viola and Jones [188], who proposed a cascade for face detection involving 32 stages. In other words, 32 stage classifiers are combined into a cascade structure. However, in the context of [188], cascading has previously existed. For instance, Rowley et al. [151] proposed the use of a fast neural network with a high false positive rate on an entire image to identify promising proposals. Positive predictions are then further passed to a more complex and slower neural network. Instead of neural networks, Serra et al. [165] used a fast linear support vector machine (SVM) followed by a more accurate polynomial SVM. According to [188] such arrangements correspond to a two-stage cascade. In this study, we refer to structures that have multiple stages as *megalith structures* and outline some of their most relevant representatives.

Megalith structure. Viola and Jones [188, 190] proposed a framework for federating cascade generation with the AdaBoost algorithm; however, their solution is not restricted to boosted classifiers. The framework essentially minimizes the expected number of weak learners to be evaluated at desired true and false positive rates. For this purpose, three hyperparameters have to be set manually: 1) the minimum true positive rate for each stage classifier, 2) the maximum false positive rate for each stage classifier, and 3) the desired false positive rate of the cascaded classifier. For each stage, the framework increases the number of weak learners, using AdaBoost, to decrease the false positive rate of the stage classifier and adjusts the classifier's threshold to increase its true positive rate. This is repeatedly performed until both requirements, the minimum true and maximum false positive rates, are satisfied for that stage. Furthermore, the framework adds new stages until the desired false positive rate of the cascaded classifier is achieved. Each new stage classifier is trained using false positives from the current cascaded classifier, a procedure known as *bootstrapping* [176]. The initial negative samples are randomly selected. The true and false positive rates of the cascaded classifier at stage S are given by $\prod_{s \leq S} R_{TP}^s$ and $\prod_{s \leq S} R_{FP}^s$, respectively, where R_{TP}^s and R_{FP}^s denote the true and false positive rates of each stage classifier.

For each stage, a high true positive rate is critical because, in the evaluation phase, improperly rejected positives cannot be recovered. The false positive rate per stage is usually less critical because the cascade false positive rate can be decreased by adding further stages; however, this is done at the cost of extra computation. At the first stage, these goals can be achieved using relatively few weak learners because the initial negative set is only randomly selected. Subsequent stages are trained using newly gathered false positives from previous stages. Consequently, subsequent stages include more difficult classification tasks and require more complex classifiers, i.e., increased number of weak learners. This way the cascade algorithm naturally creates a structure consisting of multiple classifiers that are ranked by their complexity. The cascade ROC curve can be created by moving the threshold of its last stage in the range $(-\infty, +\infty)$.

In [190], the authors demonstrated the efficiency of the cascade formation. For this, they trained a cascaded classifier consisting of ten stages with 20 decision stumps per stage and compared its performance against a *monolithic* classifier consisting of 200 decision stumps. Both classifiers were trained using the same training set. In terms of performance, both demonstrated fairly similar accuracy, with the monolithic classifier exhibiting superior performance. In terms of evaluation time, however, the cascaded classifier ran nearly ten times faster.

Megalith structures have certain conceptual shortcomings. For instance, samples can only be rejected at the end of each stage, i.e., all weak learners of a stage classifier have to be evaluated before a decision can be made. Furthermore, with an increased number of stages, the objectives that must be achieved become too complex for each individual stage classifier. For instance, to achieve an overall detection rate of 90% at a given false positive rate in a 10 stage cascade, each stage classifier must achieve a true positive rate of 99%, which results in a very difficult task for the later stages and leads to an enlarged number of weak classifiers for those stages. Moreover, two successive stage classifiers operate completely independently from each other, i.e., in case a sample passes the stage $s-1$, the cumulative sum, H_{s-1}, is discarded and does not contribute to the decision at stage s. To alleviate some of these shortcomings, some studies have focused on post-processing to optimize the cascade performance [105, 174, 108], while others suggested alternative cascade structures [13, 200, 199, 35].

Monolith structure. In [200], authors proposed a chain structure (referred to as *boosting chain*) to integrate historical knowledge, specifically, by propagating the cumulative sum (H) from prior stages as an input to the next stage. As a traditional cascade [188], boosting chain reduces the false positive rate through the addition of new stages that irreversibly reduce the true positive rate. [13] evolved this idea and accumulated information during the entire cascade evaluation. In particular, a single monolithic classifier is augmented with a rejection threshold function. This variant is called *soft cascade* and is given by

$$H_{T'}(x) = \sum_{t=1}^{T'} \alpha_t h_t(x) \geq \theta_{T'}, \qquad (2.4)$$

where $\theta_{T'}$ defines the rejection trace, i.e., the input x is rejected as soon as $H_{T'}(x) < \theta_{T'}$ for any $T' \leq T$, otherwise x is classified as positive. $\theta_{T'}$ is determined post training such that no more than a defined fraction of the positive samples is rejected.

It is worth mentioning that this monolith structure is similar to the one mentioned by Viola et al. in [188] and for $\theta_1, \ldots, \theta_{T-1} = 0$ and some θ_T, both become functionally identical. Various strategies for setting $\theta_{T'}$ have been proposed [13, 35, 206]. The constant variant of the soft cascade [35] (termed *constant soft cascade*) makes use of the fact that in practice, it is unlikely that $H_T(x) \gg 0$ while for any T', $H_{T'}(x) \ll 0$. Therefore, $\theta_{T'}$ is replaced through a global constant threshold θ^* such that $\theta_{T' \leq T} = \theta^*$. It must be emphasized that [188] and [13] use a variant of AdaBoost that combines classifiers of the type $h_t(x) \in \{0, 1\}$, i.e., $H_{T'}(x)$ rises monotonically ($\forall t\ \alpha_t > 0$ because of the definition of weak learners),

$\alpha_t h_t(x) \leq \alpha_{t+1} h_{t+1}(x)$, whereas [35] uses the variant shown in algorithm 2.2.

By adjusting θ^*, one can *calibrate* the cascade on an independent validation set. θ^* offers a trade-off between recall and speed, where recall refers to the relative number of positive samples that survive the cascade, see section 3.2. Decreasing θ^* allows more samples to pass through the classifier, which increases recall and decreases speed (and vice versa).

2.2.2 Connectionism

Connectionism denotes a set of approaches in the field of artificial intelligence that attempt to represent intellectual abilities, such as perception and cognition, by modeling mechanisms of the processing that occur in the brain. In principle, these models usually take the form of *neural networks*, i.e., networks that are composed of a large number of very simple and interconnected components called *neurons*.

2.2.2.1 Artificial neuron

An artificial neuron is the elementary computational unit in an artificial neural network. Figure 2.3 (b) illustrates the general structure of such a unit used in present neural networks. It consists of an arbitrary number of incoming edges with predetermined directions. Information flows through the directed edges and becomes aggregated, i.e., the $n + 1$ arguments pass the integration function Σ, a linear combiner that uses weighted addition to generate a single numerical value. This value is then passed into an activation function Θ. Typical choices for Θ are shown in figure 2.4. The neuron is considered as a primitive mapping function, $\Theta : \mathcal{X} \to \mathbb{R}$.

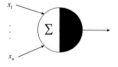

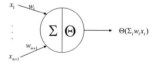

(a) MPC. AND and OR boolean operations for $\mathcal{T}_\Sigma = n$ and $\mathcal{T}_\Sigma = 1$, respectively.

(b) General structure of an artificial neuron.

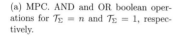

Figure 2.3: Illustration of neuron structures.

2.2.2.2   Weightless neural network - McCulloch-Pitts cell

In 1943, Warren McCulloch and Walter Pitts [115] presented a class of
neural networks to demonstrate the manner in which the brain makes complex decisions using many interconnected basic cells, the *McCulloch-Pitts
cell* (MPC for short). MPC has a simpler structure than the general representation, compare figure 2.3 (a) with figure 2.3 (b). MPC primarily
processes and outputs binary information transmitted through unweighted
edges (this is equivalent to having $w_i = 1$ in figure 2.3 (b)) so that Σ applies
a summation $\sum_i x_i$, where $x_i \in \{0, 1\}$ refers to the ith component of the
n-dimensional x. An incoming edge can either be excitatory or inhibitory,
where the latter is characterized using a small circle attached to the end
of the edge. Each neuron is assigned a fixed threshold $\mathcal{T}_\Sigma$. If the sum of
the excitements is greater or equal to the given threshold and none of the
inhibitory edges are activated, the neuron outputs 1 otherwise it outputs
0, i.e., $\Theta(\sum_i x_i) = \mathcal{H}(\sum_i x_i - \mathcal{T}_\Sigma)$ when none of the inhibitory edges are
activated with $\mathcal{H}$ being the Heaviside step function that changes discontin-
uously from 0 to 1 when $\sum_i x_i \geq \mathcal{T}_\Sigma$. This implies absolute inhibition, i.e.,
the neuron can be inactivated by one inhibition impulse. A single MPC is
capable of implementing a simple Boolean function such as AND, OR, and
NOT, see figure 2.3 (a). A composition of three units constitutes a logical basis and permits the implementation of logical functions of the type
$\{0, 1\}^n \to \{0, 1\}$ [147].

Learning with MPC. The thresholds and the connection patterns of the
network are designed to fulfill a given task; however, no training method
was presented in [115].

2.2.2.3   Weighted neural networks - perceptron

In 1958 Rosenblatt [149, 150] proposed the first algorithmically described
neural network, called *perceptron*, for the purpose of pattern recognition.
The perceptron is basically built around a single neuron, namely MPC.
The input of the perceptron is real-valued, the edges are weighted, and
the constant threshold is assigned to a free parameter called *bias* (b). The
structure can learn from its mistakes by adjusting its parameters through
a numerical algorithm. If it is a priori known that two classes are linearly separable, the perceptron algorithm is guaranteed to converge even if
its parameters are randomly initialized. The algorithm separates the two
classes by positioning a $(n - 1)$-dimensional decision hyperplane in the n-
dimensional input space. The hyperplane is given as $\sum_i w_i x_i + b = 0$. As for
MPC, the perceptron uses the Heaviside step function $\mathcal{H}$ as its activation
function. To simplify the bias representation, the input x can be extended

to a $(n + 1)$-dimensional vector $(x_1, ..., x_n, 1)$. The associated weight vector can then be given as $(w_1, ..., w_{n+1})$, whereby $w_{n+1} = b$, see figure 2.3 (b). Assume that we have two finite and linearly separable sets, $\mathcal{N}$ and $\mathcal{P}$, with $\mathcal{D} = \mathcal{P} \cup \mathcal{N}$. The perceptron algorithm finds $n + 1$ real numbers such that $\sum_i w_i x_i^{(j)} \geq 0$ if $x^{(j)} \in \mathcal{P}$ and $\sum_i w_i x_i^{(j)} < 0$ if $x^{(j)} \in \mathcal{N}$. This is done by iteratively reweighting the free parameters according to algorithm 2.3. The algorithm processes one sample at each iteration t, and therefore data points must be randomly chosen otherwise convergence is not guaranteed [72].

Learning with perceptron. Algorithm 2.3 places an initial hyperplane in the input space and moves it gradually by evaluating each sample individually into a region where both sets are separated. There exists an infinite number of decision hyperplanes that can solve such problems and algorithm 2.3 can converge to any of the decision hyperplanes. The random initialization and the way data samples are chosen make solutions not easily reproducible. Moreover, the algorithm can neither handle outliers nor nonlinearly separable data. In these two cases, it would neither find a solution within finite time nor would it approximate a correct solution [117]. To determine an optimal solution for a given distribution, in 1963, a method [183, 184] was proposed that requires from the solution to separate the data while maximizing the margin between the classes, known as *maximum margin classifier* and SVM. The margin defines the distance between the decision hyperplane and the closest data points, *support vectors*. The method specifies the position of the separator primarily according to those support vectors. In other words, other data samples do not contribute to

---

Algorithm 2.3 *Perceptron convergence algorithm*

Input:
- T number of iterations
- $\mathcal{D} = \{(x^{(i)}, y^{(i)})\}_{i \leq N}$, with $y^{(i)} \in \{0, 1\}$

Output:
- w

Initialize:
- $w^{t=1}$ is generated randomly with small numbers $\in [-1, 1]$

1: **for** $t = 1, \ldots, T$ **do**
2:     $(x, y) \in \mathcal{D}$ is selected randomly
3:     $o \leftarrow \mathcal{H}(\sum_i w_i x_i)$ evaluate unit
4:     **if** $o = y$ **then**
5:         $w^{t+1} \leftarrow w^t$
6:     **else**
7:         $w^{t+1} \leftarrow w^t + (y - o) \cdot x$
8:     **end if**
9: **end for**

---

the determination of the separator's location. The underlying intuition is that this kind of separation enables the classifier to be more confident when confronted with unseen samples, i.e., this allows the classifier to generalize better [58].

2.2.2.4 Artificial neural network

The artificial neural network, as known today, had its breakthrough based on the work of Rumelhart et al. [155, 154]. The authors proposed the use of the *error backpropagation algorithm*, that was introduced by several researchers [85, 42, 195], to broadcast the partial derivative of the output error back to the preceding layers of a *multilayer* neural network. The solution proposed by the researchers on how to train such structures is de facto fundamentally unchanged till present. In the following, we briefly review the basic composition and training routine of network structures.

Gradient descent

The error backpropagation algorithm searches for the minimum of an error function E (often also referred to as *loss* or *cost function*) in the weight space using the method of *gradient descent*. This method requires the gradient computation of the error function. Therefore, E must be continuous and differentiable. The learning task is then reduced to a minimization task and the combination of weights that achieves the lowest E is considered the solution. The form of the loss function depends on the task [84]. Two loss functions that are typically used include the *mean squared error*

$$E = \frac{1}{2}\sum_{i=1}^{N}\sum_{j=1}^{M}\left(y_j^{(i)} - o_j^{(i)}\right)^2,$$ (2.5)

which is usually used for regression tasks and the *cross entropy*

$$E = -\sum_{i=1}^{N}\sum_{j=1}^{M} y_j^{(i)} \log\left(\hat{o}_j^{(i)}\right),$$ (2.6)

which is usually used for classification tasks, i.e., y is one-hot encoded. Here, o and $\hat{o}$ denote the outputs of the models with M being their dimensions. Cross entropy makes the implicit assumption that $\hat{o}_j > 0$ and $\sum_j \hat{o}_j = 1$ for which a *softmax* function is usually used

$$\hat{o}_j = \frac{e^{o_j}}{\sum_j e^{o_j}}.$$ (2.7)

Algorithm 2.4 *Gradient descent*

Input:
- T maximum number of iterations
- ϵ minimum required error
- $\mu > 0$ learning rate

Output:
- w

Initialize:
- $t \leftarrow 0$
- w^t is initialized randomly with small numbers

1: **do**
2:    $w_p^{t+1} \leftarrow w_p^t + \Delta w_p^t$
3:    $t \leftarrow t + 1$
4: **while** $t \leq T$ or $E \geq \epsilon$

An illustration of such a model is shown in figure 2.5 (a) which employs M output neurons. Consider each neuron to be connected to all input components and let $w_{k,j}$ denote the weight that connects the kth component of the input with the jth neuron, i.e., for the ith sample $o_j^{(i)} = \Theta(\sum_{k=1}^{n+1} w_{k,j} x_k^{(i)})$, see figure 2.3 (b).

Let $E^{(i)} = (E_1^{(i)}, \ldots, E_M^{(i)})$ denote the error produced by the ith sample and $E_j^{(i)} = \frac{1}{2}\left(y_j^{(i)} - o_j^{(i)}\right)^2$ denote its jth component. When minimizing $E_j^{(i)}$, the gradient $\nabla E_j^{(i)} = (\frac{\partial E_j^{(i)}}{\partial w_{1,j}}, \ldots, \frac{\partial E_j^{(i)}}{\partial w_{n+1,j}})$ provides the direction of the steepest ascent of $E_j^{(i)}$ in the weight space $w_{k,j}$. Let $w^0 = \{w_{k,j}^0 \equiv w_p^0\}$ denote the random initialized parameter space at iteration $t = 0$, where for simplicity, the subscript $p \leq (n+1)M$ is used to index all the weights in the single layer model. Updating each weight by a portion of the gradient in the opposite direction for $0 \leq t \leq T$ iterations using

$$w_p^{t+1} \leftarrow w_p^t \underbrace{-\mu \frac{\partial E}{\partial w_p^t}}_{\Delta w_p^t}, \tag{2.8}$$

ensures convergence to a minimum, where $\nabla E = 0$, within a finite number of iterations T. In the equation above, $\mu > 0$ is called the *step size* or *learning rate*. This minimum is not guaranteed to be the global minimum (except for convex error functions) and further depends highly on the starting point (w^0). For a small value of μ, the algorithm converges within T iterations, whereas a large value of μ can cause divergence by leading the actual state uphill.

Algorithm 2.4, also known as *batch gradient descent*, summarizes this method. At each iteration, the algorithm calculates the direction of the gradient by evaluating all data samples, i.e., weight changes are accumulated over an entire presentation of the training data (an *epoch*) before being applied. When the number of training samples is high, Δw_p^t can become large, which leads to large steps and thus unstable learning behavior. A solution to address this problem involves using a small value for μ. However, this method is considered costly and slow. To accelerate the convergence, some methods have been proposed that build upon the gradient descent method [152]. Next, we outline some of those methods that we use in our study.

Stochastic and mini-batch gradient descent. To economize on the computational cost of the above-described gradient descent method, *stochastic gradient descent* approximates the true gradient by evaluating the gradient at a single sample. *mini-batch gradient descent* makes a compromise between both methods and changes weights after evaluating the gradient for a predefined number of samples. As a consequence, mini-batch gradient descent updates weights more frequently than gradient descent and makes a more robust approximation of the true gradient by considering more than one training sample.

Momentum. Momentum [137] accelerates convergence by determining the next update as a linear combination of the gradient and the previous update

$$\Delta w_p^t = -\mu \frac{\partial E}{\partial w_p^t} + \alpha \Delta w_p^{t-1}, \qquad (2.9)$$

where $0 \leq \alpha < 1$ is the momentum rate, typically $\alpha \approx 0.9$. The momentum term causes a kind of smoothness in the convergence behavior and reinforces updates in directions where gradients point more frequently.

Weight decay. Weight decay [95] is an approach that limits the complexity of a neural network to prevent overfitting. The approach does that by introducing an additional term to the error function

$$E^+ = E + \frac{\lambda}{2} \sum_p w_p^2, \qquad (2.10)$$

that limits the growth of the weights and encourages a decay toward zero. Applying gradient descent to E^+ leads to

$$\Delta w_p^t = -\mu \frac{\partial E}{\partial w_p^t} - \underbrace{\mu \lambda w_p^t}_{\text{decay term}}, \qquad (2.11)$$

where λ is a small positive constant to choose beforehand, typically $\lambda \approx$

0.0005.

Adam. Adaptive moment estimation (Adam) [88] changes the weights based on exponential moving averages of the gradient (m^t) and the squared gradient (v^t)

$$\Delta w_p^t = -\mu \frac{\hat{m}_p^t}{\sqrt{\hat{v}_p^t} + \epsilon},\tag{2.12}$$

with

$$\hat{m}_p^t = \underbrace{\frac{m_p^t}{1 - \beta_1^t}}_{\text{bias-corrected}}, \quad \underbrace{m_p^t = \beta_1 m_p^{t-1} + (1 - \beta_1)(\frac{\partial E}{\partial w_p^t})}_{\text{first moment estimate}},$$

$$\hat{v}_p^t = \underbrace{\frac{v_p^t}{1 - \beta_2^t}}_{\text{bias-corrected}}, \quad \underbrace{v_p^t = \beta_2 v_p^{t-1} + (1 - \beta_2)(\frac{\partial E}{\partial w_p^t})^2}_{\text{second moment estimate}},$$

where the hyperparameters β_1, $\beta_2 \in [0, 1)$ determine the exponential decay rates of the moving averages, with β^t denoting β to the power of t. m^t and v^t are estimates of the first (the mean) and the second (the uncentered variance) moments of the gradient. Adam uses bias-corrected estimates $\hat{m}^t$ and $\hat{v}^t$ for updating the weights because m^t and v^t are initialized as zeros and therefore biased toward zero. In various tasks, Adam has demonstrated suitability for problems that are large in terms of data and/or parameters [88], where its hyperparameters require little tuning and are typically initialized with $\alpha = 0.001$, $\beta_1 = 0.9$, $\beta_2 = 0.999$ and $\epsilon = 10^{-8}$.

Activation functions

Requiring E to be differentiable consequently means that the activation functions of the neurons must share this property. In the early years of neural networks, the *sigmoid*: $\mathbb{R} \to (0, 1)$ and *hyperbolic tangent*: $\mathbb{R} \to (-1, 1)$ were predominantly used in such structures. Both functions are shown in figure 2.4 (a) and (b). The output of both functions reach asymptotically some extremes. Near the extremes, the neuron becomes *saturated*, i.e., the inputs or the weights must have a large magnitude to change the neuron's behavior in those regions. Moreover, because of the limits of the functions, the derivative in those regions becomes very small and therefore decelerates learning via gradient descent. This has shown to be unfavorable for training *deep* networks, therefore, the *rectified linear unit, ReLU*: $\mathbb{R} \to [0, \infty)$ shown in figure 2.4 (c), was proposed for the use in deeper structures [69]. The derivative at $\frac{d}{dx}\text{ReLU}(x = 0)$ does not exist, for which, an arbitrary value is technically assigned to that region, usually 0. This makes the error backpropagation algorithm applicable to this activation function. One drawback of ReLU is that learning happens only for $\text{ReLU}(x) > 0$ because only then

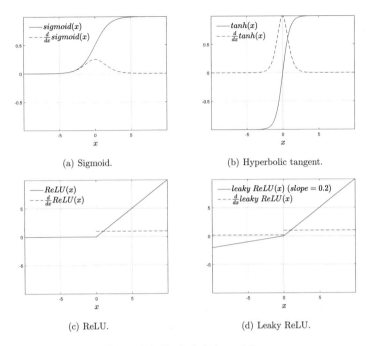

(a) Sigmoid.

(b) Hyperbolic tangent.

(c) ReLU.

(d) Leaky ReLU.

Figure 2.4: Typical choices of Θ.

$\frac{d}{dx}$ReLU$(x)\neq 0$. A common solution to permitting more non-zero derivatives includes the use of *leaky ReLU*$= max(x \cdot$ slope$, x) \in (-\infty, +\infty)$ [110], shown in figure 2.4 (d). The slope value is usually < 1 and for slope$= 0$ leaky ReLU becomes identical to ReLU.

The network function

Figure 2.5 (b) illustrates a network, more concretely, a two-layered *fully connected* (FC), *feedforward* model, where FC implies that each neuron is connected with all neurons from its previous layers, and feedforward implies that the connections between the neurons do not form a cycle. The network uses the extended input and weight notations, see figure 2.3 (b), has K *hidden* and M output neurons, where the term hidden denotes neurons or layers that do not belong to the input layer (which is not counted as a layer) nor to the output layer. The weight $w_{k,j}^l$ connects the kth neuron from layer $l-1$ (for $l > 1$) or the kth input component (for $l = 1$) with the jth neuron of layer $l \leq L$, where $L = 2$ is the total number of layers in the network depicted in figure 2.5 (b). The network is a computational graph and an

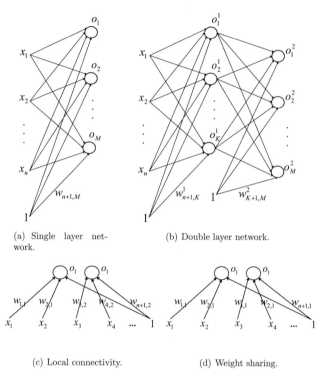

(a) Single layer net-   (b) Double layer network.
work.

(c) Local connectivity.   (d) Weight sharing.

Figure 2.5: Illustration of different network structures.

implementation of a particular composite function

$$o^L(x; w^1, \dots, w^L) = \Theta^L(\ \dots\ \Theta^2(\Theta^1(x; w^1); w^2)\ \dots\ ; w^L), \qquad (2.13)$$

that maps an input x into the output $o^L = (o_1^L, \dots, o_M^L)$. The error back-propagation applies the chain rule to propagate the error, recursively, from the top (the cost function) back to the lowest layer of the network. Therefore, training operates in three phases:

- Propagate activity forward: The network receives samples and the activations of the respective neurons are propagated forward, from the input to the output neurons. This is known as *forward pass*.

- Propagate gradients backward: The error is computed according to the cost function and its derivatives are propagated back layer by layer starting from the output to the input neurons. This is known as *backward pass.*

- Change weights: Update each weight via the gradient descent method following algorithm 2.4 or any of the related methods.

This training procedure applies also to structures with $L > 2$.

2.2.2.5 Convolutional neural network

The convolutional neural network (CNN) [101] introduces three new concepts into network structures to process images more efficiently. These new concepts include *local connectivity*, *weight sharing*, and *pooling*, see figure 2.5 (c) and (d). In the convolution layer, the neurons are organized into planes, where each plane is called a feature map. Each layer can contain multiple feature maps. In a feature map, neurons are *locally* connected, i.e., each neuron takes inputs only from a small subregion of the image or its previous layer. This subregion is also known as a *receptive field* because stimuli from that region modify the activation of the neuron. In images, nearby pixels are typically more correlated than more distant pixels. Using local connectivity, each neuron learns to extract *local* features that depend only on the small subregion of the image. In a feature map, all neurons are constrained to *share* the same weights. For instance, a feature map can consist of 200 neurons arranged in a 20×10 grid with each neuron receiving inputs from a 5×5 subregion of the image, which leads to 26 trainable weights (25 for the weights plus bias). Considering the neuron to be a local feature detector, weight sharing means searching for the same pattern at all subregions in the image. Because of weight sharing, the preactivation (see figure 2.3 (b), linear combiner) of the feature map resembles a convolution of the image pixel intensities with a kernel comprising the weight parameters, hence the name of the network. Note that some implementations use cross-correlation instead of convolution [186]. We use the following equation to formulate the activation of the feature map

$$o^l(x,y) = \Theta \left(\sum_{x',y',c'} w^l(x',y',c')o^{l-1}(\delta^w x + x', \delta^h y + y', c') + b^l \right), \quad (2.14)$$

where $x', y', c' \in \mathbb{N}$ defines the subregion for each neuron, and $\delta^w, \delta^h \in \mathbb{N}^+$ are the spatial strides of the kernel in horizontal and vertical directions, respectively. $\delta^w, \delta^h > 1$ subsample the representation, i.e., reduce its spatial size in horizontal and vertical dimensions. The final concept, pooling, combines the outputs of neighboring neurons at one layer into a single neuron input in the next layer using either *max pooling*,

$$o^l(x, y, c) = \max_{x', y'} o^{l-1}(\delta^w x + x', \delta^h y + y', c), \qquad (2.15)$$

or *average pooling*

$$o^l(x, y, c) = \frac{1}{S^2} \sum_{x', y'=0}^{S-1} o^{l-1}(\delta^w x + x', \delta^h y + y', c), \qquad (2.16)$$

where the parameter S is predefined. A typical CNN structure consists of a repetition of convolution and pooling layers followed by some FC layers [93].

2.3   Proposal generation

The aim of object detection methods is to determine whether an object exists in an image, and if so the location of the object in the image. Over the past decade, a dominant approach for solving this problem has been an exhaustive search over *quasi* every location and scale in an image [190, 30, 49]. This approach frames the object detection as an iterative classification problem. In the following, we briefly describe this approach known as the *sliding window*, which relies on two main hyperparameters:

- Spatial stride: A detection window (also known as *model template*) with a fixed size is usually placed at different locations in the image. The distance between the neighboring locations is either δ^w in the horizontal or δ^h in the vertical direction with $\delta^w, \delta^h \ll W^d, H^d$, where H^d and W^d are the height and width of the detection window. The parameters δ^w and δ^h, usually $\delta^w = \delta^h$, are referred to as *spatial strides*.

- Scale stride: Because of the fixed size of the model template, it only captures instances of a specific scale. There are different approaches for presenting proposals of different sizes to the classifier. We postpone the discussion about these approaches to section 2.4.0.4 since such a discussion is closely related to another one regarding the properties of the feature representation which is presented later. Most methods, however, that employ the sliding window search usually employ an *image pyramid* for that purpose. Therefore, the input image is typically resampled with δ^s scales per octave — an octave is the interval between one scale and another with half or double its value — and the detection window is applied to each resampled image. The parameter δ^s is referred to as *scale stride*.

The sliding window approach (depending on the detection task, spatial and scale strides) results in numerous patches/proposals that must be evaluated by the classifier. These proposals are *class-aware* because the fixed size and aspect ratio of the window is set to fit the object of interest. The part-based object localization method of Felzenszwalb et al. [49] is related to this approach. The difference is that it searches for objects and object parts. Lampert et al. [99] proposed the use of a quality function that guides the search. Searching for an optimal window within the image then corresponds to searching for the global maximum of this function and is efficiently performed using the branch and bound technique. This method alleviates the constraints of using a window of fixed size and aspect ratio, while at the same time the number of locations visited is reduced.

Advancements in deep networks have stressed the need for other, more efficient sampling methods because of the expensive feature creation and classifier evaluation of deep models. To direct computational effort to focus only on the most salient and distinctive locations in an image, some methods apply an inexpensive detection method to preselect promising, class-aware proposals. These methods are discussed in more detail in chapter 4.

A different line of work has resulted in region proposal algorithms that consider an image as the input and output bounding boxes corresponding to all patches that are most likely to be objects. In the following, we outline some of the most relevant proposal generation methods, a comprehensive discussion can be found in [77, 76, 81]. Most discussed methods generate proposals that are *class-agnostic*, which practically means that these algorithms segment an image in foreground objects and backgrounds, based on some criteria such as color, edges, texture, etc. Region proposal methods are required to have a relatively high recall. The precision of such methods, however, is less critical because the number of generated proposals is usually small relative to the total candidates typically considered by sliding window approaches.

Zitnick et al. [215] proposed the use of *Edge Boxes*, a method that scores *objectness* primarily based on the number of contours wholly enclosed in an image patch. Authors in [141, 1] used multiple image cues, beside edges, to measure characteristics of objects. *Selective Search* [181] has been successfully used by recent detection methods [68, 181]. This is because of its relatively fast runtime (seconds per image) and high recall. Selective Search computes multiple hierarchical segmentations based on superpixels [51] and places bounding boxes around each segment. *Regions with CNN features* (R-CNN) [68] is the most prominent CNN-based detection method that uses region proposals from [181]. R-CNN is basically divided into four sub-processes: 1) extracting a given number of proposals (foreground objects) using [181], 2) creating CNN features of the extracted proposals, 3) applying a classifier on the created CNN features, and 4) using post-processing to refine bounding boxes. Two of the direct descendants of this method, fast R-CNN [67] and faster R-CNN [144], use a more efficient mechanism to process the proposals. Instead of repeatedly creating CNN features for each RGB proposal, both pass the entire input image to the CNN and evaluate CNN features from the last convolution layer that correspond to each proposal's location. The main difference between fast and faster R-CNN is that the latter introduces a region proposal network (RPN) for scoring objectiveness which shares convolutions with the classifier network.

To accelerate CNN-based detection methods, authors have suggested using a single feedforward network to address proposal generation and classification in one step. For this purpose, Lenc et al. [103] used a data-driven sampling strategy. An image-independent list of candidate regions is created by studying the distribution of the object locations in the dataset. Authors demonstrated that using such a list, a bounding box regressor can recover sufficiently accurate bounding box locations. Redmon et al. [142] used a similar idea and framed the object detection as a regression problem to spatially separated bounding boxes. For this, they divided the input image into a $S \times S$ grid and trained a CNN that predicts B bounding boxes for each cell in this grid (each bounding box has four coordinates), a confidence score for each bounding box, and the probabilities for C classes, i.e., for each image, the CNN outputs a three-dimensional tensor of the size $S \times S \times (B \cdot 5 + C)$ that encodes detection coordinates, objectiveness, and class probabilities. Yoo et al. [204] used a similar strategy in an iterative manner. For this purpose, the CNN receives an input and outputs, besides class prediction, the ways of moving the top-left and bottom-right corners of the input patch toward the object location. Cropping the input accordingly and presenting the new patch to the CNN repeatedly leads to the accurate location of the object.

2.4   Feature representation

In modern deep learning, the distinction between feature creation functions and classifiers disappears. Neural networks directly map input images into classification predictions, and therefore, effectively function as a feature extractor and classifier at the same time. The networks automatically create representations in the intermediate layers and therefore do not need hand-designed representations. Overall, the trend results in simpler, larger, and deeper network-based methods that fulfill different tasks at once [142, 74]. In pedestrian detection, however, many state-of-the-art methods either rely on detection pipelines that use handcrafted features [146, 19] or complement CNN representations with such features [20, 202, 19]. In the following, three of the most relevant feature creation methods are discussed in detail.

2.4.0.1 Histogram of oriented gradients

Dalal and Triggs [30] proposed a sophisticated pipeline to extract local shape-related information referred to as *histogram of oriented gradients* (HOG). HOG counts occurrences of gradient orientations in localized portions of an image. This is similar to methods like edge orientation histograms [53, 54], scale-invariant feature transform [106], and shape contexts [7]. A fundamental difference between the mentioned methods and HOG is that Dalal and Triggs used a dense grid of uniformly spaced cells and overlapping local contrast normalization. HOG divides the input image into small connected cells. For each cell, a local one-dimensional (1-D) histogram of the edge directions is computed over the pixels of the cell. For improved accuracy, these local histograms are normalized by an intensity value calculated across several neighboring cells, so-called *blocks*. This normalization results in better invariance to changes in illumination, shadowing, and foreground-background contrast. The feature vector is the concatenation of all histograms computed from the input patch. The underlying principle behind HOG is that local object appearance and shape within an image can be described rather well by their edge directions and intensities, even without precise knowledge of their corresponding locations. The authors of [30] performed exhaustive experiments to demonstrate the contribution of various components of their feature extraction method to the final detection accuracy. In the following, we review each component, the implementation choices, and the most important insights presented in [30].

Gamma or color normalization. Gamma/color normalization usually serves as a pre-processing step before feature creation. It helps to make features more robust to variations in illumination.

Gradient computation. The image gradients

$$\nabla I(x,y) = (\frac{\partial I}{\partial x}, \frac{\partial I}{\partial y}) \equiv (I_x, I_y),\tag{2.17}$$

are computed using local differences. For this, there exist several discrete derivative masks:

- 1-D centered: $\begin{pmatrix} -1 & 0 & 1 \end{pmatrix}$, $\begin{pmatrix} -1 & 0 & 1 \end{pmatrix}^T$
- 1-D uncentered: $\begin{pmatrix} -1 & 1 \end{pmatrix}$, $\begin{pmatrix} -1 & 1 \end{pmatrix}^T$
- 1-D cubic corrected: $\begin{pmatrix} 1 & -8 & 0 & 8 & -1 \end{pmatrix}$, $\begin{pmatrix} 1 & -8 & 0 & 8 & -1 \end{pmatrix}^T$

- 2-D Sobel mask: $\begin{pmatrix} 1 & 0 & -1 \\ 2 & 0 & -2 \\ 1 & 0 & -1 \end{pmatrix}$, $\begin{pmatrix} 1 & 0 & -1 \\ 2 & 0 & -2 \\ 1 & 0 & -1 \end{pmatrix}^T$
- 2-D diagonal: $\begin{pmatrix} 0 & 1 \\ -1 & 0 \end{pmatrix}$, $\begin{pmatrix} 0 & 1 \\ -1 & 0 \end{pmatrix}^T$

Before the gradient calculation, Gaussian smoothing is often applied to reduce the level of noise in an image, which preserves boundaries and edges.

Counting local gradient orientations. For each discrete pixel position of the input image $I(x, y)$, the orientation, $O(x, y)$, and gradient magnitude, $M(x, y)$, are defined as

$$O(x,y) = \arctan\left(\frac{I_y(x,y)}{I_x(x,y)}\right), \tag{2.18}$$

$$M(x,y) = \sqrt{I_x(x,y)^2 + I_y(x,y)^2}. \tag{2.19}$$

To calculate the 1-D histogram within the cell, β, each orientation $O(x, y)$ is assigned to one of the Q bins of the histogram, therefore $O(x, y)$ first needs to be quantized. Let $O_q(x, y)$ denote the quantized $O(x, y)$ so that $O_q(x, y) \in \{q_1, \ldots, q_Q\}$. To limit aliasing, votes are (bilinearly) interpolated between the neighboring histogram bins. The votes of the histogram are further weighted by a function of the gradient magnitude such that for each bin q_i of the histogram h, the frequency h_{q_i} in β is given by

$$h_{q_i} = \sum_{x,y \in \beta} W(x,y)\mathbb{1}[O_q(x,y) = q_i], \tag{2.20}$$

with

$$W \in \{M, M^2, \sqrt{M}, min(\delta, M)\}, \tag{2.21}$$

where $\delta \leq 0.2$ is the clipping constant. The Q bins of the cell histogram are evenly spaced either over $0° - 180°$ (unsigned gradients) or $0° - 360°$ (signed gradients). Cells can either group a rectangular or radial formation of pixels.

Block building and normalization. The magnitudes of the gradients can vary because of the variation in illumination. Dalal and Triggs alleviated this by employing local contrast normalization. For that, neighboring cells build a block, blocks may overlap with neighboring blocks, and each block is normalized individually. In other words, each cell's response may contribute multiple times to the final feature vector as it may contribute to multiple blocks. There are several schemes for block normalization. Let v denote the 1-D unnormalized descriptor vector of a given block and $|v|_p = \sum_i(|v_i|^p)^{\frac{1}{p}}$. The schemes include:

- L2-norm: $v \leftarrow \frac{v}{\sqrt{|v|_2^2 + \epsilon^2}}$
- L2-Hys: a) L2-norm, b) $v_i \leftarrow min(\delta, v_i)$, c) L2-norm

- L1-norm: $v \leftarrow \frac{v}{|v|_1 + \epsilon}$
- L1-sqrt: $v \leftarrow \sqrt{\frac{v}{|v|_1 + \epsilon}}$

with ϵ being a small regularization constant that is needed because the operation also applies over empty patches. As mentioned previously, for the cell construction, two block geometries can be applied: rectangular (R-HOG) or circular blocks (C-HOG).

Evaluation and insights. The experiments in [30] demonstrated that using RGB or LAB images results in comparable performances and is superior to using grayscale images. Furthermore, a global normalization of the input signal, as a pre-processing step, has little effect on the results because of the subsequent block normalization. For gradient calculation, smoothing and using larger masks have been proved to be unfavorable. The best performance was achieved with the 1-D centered derivative mask without smoothing. A detailed evaluation further demonstrated that the detector cues mainly on the silhouette contours of local shapes such as head, shoulder, and feet of the pedestrian. Therefore, the cell size must spatially fit those characteristics. The best reported performance in pedestrian detection was achieved using 6×6 [pixel] square cells, 3×3 [cell] R-HOG blocks with overlap, 9 orientation bins of histograms spaced between $0° - 180°$ and weighted with $W(x, y) = M(x, y)$. Although the overlapping seems to be redundant, it significantly improved the normalization quality and feature ability to accentuate foreground contours from cluttered backgrounds. The block normalization, as the study showed, is an essential step in the pipeline and significantly improves the quality of the features. While the above-mentioned normalization schemes resulted in comparable performances, the $L1 - norm$ achieved slightly inferior performance.

The local description of characteristic structures makes HOG insensitive to geometric translations to a controllable degree, and the strong local normalization largely improves feature insensitivity to photometric translations. Because of these features, HOG has been successfully adopted for numerous object detection tasks and became one of the most used features in the PASCAL object challenges [138].

2.4.0.2 Integral channel features

Porikli [139] proposed an efficient computation of rectangular local histograms using integral image representations [190], *integral histograms*. The idea is that, given a representation

$$h_{q_i}(x, y) = W(x, y) \mathbb{1}[O_q(x, y) = q_i], \tag{2.22}$$

the number of orientations equal to q_i in any region of $O_q(x, y)$ is the sum over the same region in $h_{q_i}(x, y)$. Therefore, using an integral image representation of h_{q_i} allows for any local histogram to be quickly computed with few arithmetic operations [190]. Zhu et al. [214] used a similar approach to approximate HOG features likewise [100, 211].

Dollár et al. [37], inspired by [180], introduced the *integral channel features* (ICF). ICF uses a series of translationally invariant linear and nonlinear image representations, called *channels*. The integral image representations of the channels, referred to as *integral channels*, are created to efficiently compute different types of features such as local sums, histograms, Haar-like features [133, 190] and their various generalizations [38]. ICF couples diverse information from the use of different channel types: 3 LUV color, 1 gradient magnitude, and 6 gradient orientation channels.

2.4.0.3   Aggregated channel features

Dollár et al. [34] introduced the *aggregated channel features* (ACF), that significantly simplifies the feature creation of ICF while using the same channels. For this, ACF divides all channels into equally sized, not overlapping square blocks and aggregates channel values in each block. These aggregated values correspond to the first-order features (local sums) used in ICF. In doing so, each value in the resulting lower resolution channels can directly be used as a feature, i.e., features are single value lookups.

2.4.0.4   Features and detection schemes

Translation invariance. The ICF representation uses a set of translationally invariant channels. Let Ω denote a channel (C) generation function, $C = \Omega(I)$. For any I_1 and I_2 related by a translation, $C_1 = \Omega(I_1)$ and $C_2 = \Omega(I_2)$ must be related by the same translation. This attribute of the representation is essential for fast detector evaluation because it allows the one-time creation of C for the entire image I rather than in a patch-wise manner for overlapping proposal locations. This means that for a multi-scale evaluation, a dense image pyramid is created and for each stage of the pyramid, translation-invariant channels are created, constituting a *channel pyramid*. A model trained for a specific scale is then applied to each stage of the channel pyramid, see figure 2.6 (b).

Scale invariance. ICF features are not scale-invariant, that is $\sum_{x,y} M(x, y) \neq \sum_{x,y} M_s(x, y)$, where M and M_s are the gradient magnitude channels of the given images I and I_s, respectively, with I_s being

I resampled by a factor *s*. Given translation-invariant channels and scale-invariant features, fast multiscale detection can be performed through the construction of a *classifier pyramid* as in [188], i.e., scaling is achieved by scaling the model itself, see figure 2.6 (a). Note that, figure 2.6 (a) also represents a naive approach [8] in which a number of scale-specific models are applied to a single-scale channel. Utilizing integral image representation, the Haar-like features can be created with a fixed number of operations at any scale and location. To create these features at a different spatial position, the feature creation function must be shifted accordingly on the integral image representation. To create these features at a different scale, the function must be shifted and the dimension adjusted accordingly. Dollár et al. [36, 34] proposed an approach to make channel features partially, for

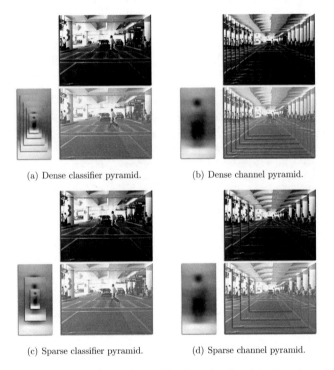

(a) Dense classifier pyramid.   (b) Dense channel pyramid.

(c) Sparse classifier pyramid.   (d) Sparse channel pyramid.

Figure 2.6: Different schemes for multiscale pedestrian detection. In each of the subfigures, the first row contains either the image or the image pyramid. The second row (left) contains either the model template or the classifier pyramid. The second row (right) contains either the channel or the channel pyramid.

nearby scales, insensitive to small changes in scale, which is referred to as *channel approximation*. In the following, we briefly describe this approach.

For nearby scales and for numerous shift-invariant channel functions, the relation $\frac{\sum_{x,y} \Omega(I)(x,y)}{\sum_{x,y} \Omega(I_s)(x,y)} \approx \mu$ is fulfilled, irrespective of the underlying interpolation method used for resampling and the size of I. In other words, the sums over any $\Omega(I)$ and $\Omega(I_s)$ differ by a multiplicative constant for a given scale s and the overall relationship, independent of the scale at which the image was captured, follows the power law. Let $f_\Omega(I_s) = \frac{1}{H^{C_s}W^{C_s}} \sum_{x,y} C_s(x,y)$ denote the global mean on the channel C_s, where H^{C_s} and W^{C_s} denote the height and width of C_s, respectively. Assuming $f_\Omega \neq 0$, for each given channel function, λ_Ω in

$$f_\Omega(I_{s_1})/f_\Omega(I_{s_2}) = (s_1/s_2)^{-\lambda_\Omega} + \epsilon, \qquad (2.23)$$

can be approximated such that $E[\epsilon] \approx 0$, where ϵ is the deviation from the power law. For $s_2 = 1$ (i.e., $I_{s_2} = I$) and $s_1 = s$, the above expression can be simplified to $f_\Omega(I_s) \approx f_\Omega(I) \cdot s^{-\lambda_\Omega}$. Benenson et al. [8] incorporated this idea and proposed to create a sparse classifier pyramid and used the above approximation to *rescale* the classifier for the intermediate scales and consequently mimic a dense classifier pyramid, see figure 2.6 (c). Dollár et al. [34] proposed to create a sparse channel pyramid with stages at each octave and incorporate the above to approximate feature responses at intermediate scales, see figure2.6 (d). Since the quality of the approximation degrades with increasing values of s, both approaches consider only the nearest scales for the approximation. Both approaches reduce the number of times images are resized and features computed, which reduces the total detection time.

2.5 Clustering the detections

Current object detectors are usually incapable of producing exactly one detection bounding box per instance. A majority of classifiers are, to a certain grade, insensitive to scale and translation, i.e., multiple neighboring regions trigger the classifier, which results in detection bounding boxes with relatively high scores near the real location of the object of interest. The number and density of these outputs (also referred to as *score map*) depend, among other aspects, on the sampling strategy, for instance, the spatial and scale strides of the sliding window approach. The number of responses in such score maps is only loosely correlated with the real number of the objects of interest in the image. Therefore, this number is not sufficient for understanding the content of the image and detectors usually employ a post-processing step to eliminate redundant boxes. The de facto standard algorithm for this task that has been used for several generations of detectors, from Viola et al. [190], Dalal and Triggs [30] to the current state-of-the-art R-CNN family [68, 67, 144], is the *non-maximum suppression* (NMS) algorithm. The goal of the NMS algorithm is to retain the local maximum in the score map while suppressing nearby responses with less confidence in a group of detections. This maximum, ideally, belongs to the corresponding detection window that contains the object of interest. Therefore, NMS takes a set of detection bounding boxes $\mathcal{B}$ and the corresponding scores $\mathcal{S}$ as an input. The algorithm then exhaustively compares each pair in $\mathcal{B}$ and suppresses the ones with smaller confidence based on an overlap threshold $\mathcal{T}_o$. The NMS procedure is $\mathcal{O}(N_{\mathcal{B}}^2)$ in the number of bounding boxes.

There are different variants of the NMS procedure, for instance, Viola and Jones in [188] used NMS for face detection to partition bounding boxes into disjointed subsets based on their area of overlap. Instead of suppressing, the

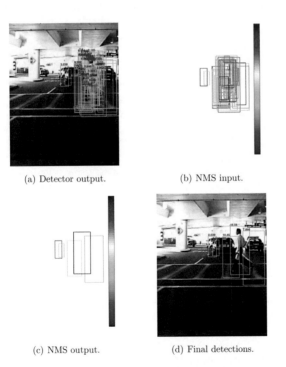

(a) Detector output.   (b) NMS input.

(c) NMS output.   (d) Final detections.

Figure 2.7: Illustration of the non-maximum suppression procedure.

Algorithm 2.5 *Non-maximum suppression*

Input:
- $\{(b_i, s_i)\}_{i \leq N_B}$, bounding boxes and their corresponding scores
- $f_o(b_i, b_j)$ overlap function of b_i and b_j
- $\mathcal{T}_o$ overlap threshold

Output:
- $\mathcal{B}^d$ set of final bounding boxes

Initialize:
- $\mathcal{B}^d \leftarrow \emptyset$

1: **while** $\mathcal{B} \neq \emptyset$ **do**
2:    $i \leftarrow argmax\ \mathcal{S}$
3:    $\mathcal{B}^d \leftarrow \mathcal{B}^d \cup b_i$
4:    $\mathcal{B} \leftarrow \mathcal{B} \setminus b_i$
5:    $\mathcal{S} \leftarrow \mathcal{S} \setminus s_i$
6:    **for** b_j *in* $\mathcal{B}$ **do**
7:        **if** $f_o(b_i, b_j) > \mathcal{T}_o$ **then**
8:            $\mathcal{B} \leftarrow \mathcal{B} \setminus b_j$
9:            $\mathcal{S} \leftarrow \mathcal{S} \setminus s_j$
10:       **end if**
11:   **end for**
12: **end while**

---

coordinates of bounding boxes within each subset are averaged to return a final box. The most common NMS approach for object detection does not consider discarded boxes [30, 49, 37] (performs in a greedy fashion), i.e., once a bounding box is suppressed, it can no longer suppress other bounding boxes.

Algorithm 2.5 summarizes the NMS procedure. The function $f_o(b_i, b_j)$ is given as

$$f_o(b_i, b_j) \in \left\{ \frac{area(b_i \cap b_j)}{area(b_i \cup b_j)}, \frac{area(b_i \cap b_j)}{min(area(b_i), area(b_j))} \right\}, \qquad (2.24)$$

where the first variant is known as the *intersection over union* (IoU). Figure 2.7 illustrates the NMS procedure. (a) presents the actual output of a pedestrian detector where the red numbers above the bounding boxes are their corresponding scores. (b) illustrates the input to the NMS procedure. The color of each box represents the score of that box, i.e., red and blue colors represent high and low confidences, respectively. Red boxes are first considered by the algorithm. (c) shows the output $\mathcal{B}^d$ of algorithm 2.5. Similarly, the colors represent the corresponding scores. Finally, (d) presents the resulting score map on the input image.

Because of the hand-designed components, the NMS strategy must be adapted to the detection task. The choice of the threshold $\mathcal{T}_o$ poses a trade-off between precision and recall. If $\mathcal{T}_o$ is too large (too strict), not enough surrounding boxes are suppressed and false positives with relatively high scores which surround the object of interest remain and thus the precision suffers. If $\mathcal{T}_o$ is too low (too loose), then multiple true positives can be merged together and the recall suffers. Moreover, algorithm 2.5 and figure 2.7 demonstrate that NMS fails to remove false positives that do not have any/enough overlap with true positives, irrespective of the choice $f_o(b_i, b_j)$ and $\mathcal{T}_o$.

Some methods have been proposed as alternatives to the NMS procedure. In [79, 78], a CNN is trained to suppress false positives. For this, the CNN in [79] receives boxes and scores and operates concurrently on the boxes and scores, as opposed to NMS. Such methods do not include a grid search on the validation set for choosing the right parameters and can learn to adapt to the distribution of data. Bodla et al. [12] suggest the framing of NMS as a rescoring procedure, termed as *Soft-NMS*. Assuming that $\mathcal{S} \in \mathbb{R}^+$, line 9 in algorithm 2.5 can be rewritten as $s_j \leftarrow s_j \cdot 0$. The authors suggest the use of either a linear,

$$s_j \leftarrow s_j(1 - IoU(b_i, b_j)), \tag{2.25}$$

or a Gaussian,

$$s_j \leftarrow s_j \exp^{-\frac{IoU(b_i, b_j)^2}{\sigma}}, \tag{2.26}$$

rescoring function, where $\sigma \approx 0.5$ remains to be manually set. Both functions reduce the detection scores with respect to the overlap with neighboring bounding boxes.

Chapter 3

Datasets and Methodology

3.1 Introduction

This chapter introduces the datasets and benchmarks used throughout this study. In section 3.2, we describe the datasets used for pedestrian detection, dataset evaluation protocols, and the most relevant features of each dataset. In section 3.3, we briefly describe the traffic sign samples and the camera sensor providing this information.

3.2 Pedestrian detection

The pedestrian datasets, Caltech [39] and Kitti [60], that we use in this study provide recordings from a moving platform and a per-image evaluation protocol. A comprehensive overview of other relevant datasets and their properties can be found in [91, 40, 60].

Any multiscale detection dataset typically consists of a given number of images and their corresponding annotations. Each annotation contains a set of ground truths, each giving the location (given as a bounding box) and the corresponding label of a relevant object in the image.

Stationary and mobile platforms. These images can be acquired from photographs [30], surveillance videos [197], or mobile recording platforms such as a vehicle [196]. The first two approaches are typically influenced by bias. As photographs are frequently manually selected, they suffer from a selection bias; surveillance videos suffer from a restricted background. Therefore, the last approach is preferred for evaluating pedestrian detection methods. Continuous recording from a moving platform eliminates both disadvantages identified above.

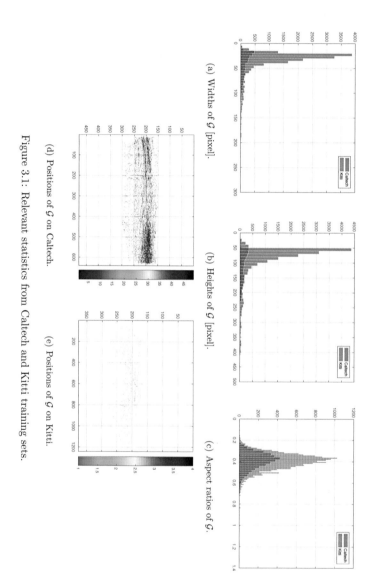

(a) Widths of $\mathcal{G}$ [pixel].

(b) Heights of $\mathcal{G}$ [pixel].

(c) Aspect ratios of $\mathcal{G}$.

(d) Positions of $\mathcal{G}$ on Caltech.

(e) Positions of $\mathcal{G}$ on Kitti.

Figure 3.1: Relevant statistics from Caltech and Kitti training sets.

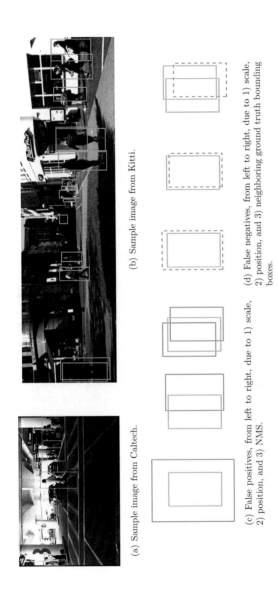

(a) Sample image from Caltech.

(b) Sample image from Kitti.

(c) False positives, from left to right, due to 1) scale, 2) position, and 3) NMS.

(d) False negatives, from left to right, due to 1) scale, 2) position, and 3) neighboring ground truth bounding boxes.

Figure 3.2: (a) and (b) show sample images from the Caltech and Kitti training sets. (c) and (d) show scenarios that may not be explicitly examinable in a per-image evaluation. Green boxes represent the ground truth bounding boxes. Solid and dashed blue boxes represent the detection bounding boxes for positive and negative predictions, respectively.

Per-window and per-image evaluation. Datasets can contain only pedestrian windows and are therefore primarily suited to train and test binary classification algorithms. To evaluate the algorithms, their per-window performances on cropped positive and negative image windows are measured. Such benchmarks isolate the classifier performance from the overall detection system. Conversely, datasets that contain pedestrians in their original contexts allow designing and testing of full-image detection systems. A detector is not only required to report the presence of an object of interest but also to determine its location in an image. Based on the detection methodology, the detector includes, in addition to the classifier, strategies to sample from the input image, such as spatial and scale strides, and NMS, see sections 2.3 and 2.5. These choices influence the full image performance, and as discussed in [40], in practice, the per-window and per-image evaluations can only be weakly correlated. Figure 3.2 (c) and (d) display scenarios that can be considered in a per-image evaluation, yet due to the evaluation protocol, may not be examinable in a per-window evaluation.

3.2.1   Caltech benchmark

The Caltech pedestrian dataset [39], proposed in 2009, is considered as one of the largest datasets of its kind to date. It includes 350000 bounding boxes labeling 2300 unique pedestrians in 250000 frames with a resolution of 480×640 (height $\times$ width). The frames were captured in the USA, over 11 sessions, from a vehicle driving in regular traffic in an urban environment. The data is divided approximately in half, to create a standard protocol for training and evaluating detection methods. Therefore, authors [39, 40] set aside six sessions (S0-S5) for training and five sessions (S6-S10) for testing. The standard training set (every 30th frame is considered) contains 4250 frames with nearly 2000 annotated pedestrians. The test set includes 4024 frames with 1000 pedestrians. Because all of the frames in the videos are fully annotated, the size of the training set can be easily increased by decreasing the sampling interval (SI). A setting that is frequently used in recent studies is referred to as *Caltech10×*, see [210, 80, 209]. This setting resamples the training set using SI=3; this increases the number of training samples (linear growth in the number of pedestrians) by a factor of ten.

Annotations. Caltech contains three types of instances that have corresponding bounding boxes labeled as:

- 'Person' — individual pedestrian
- 'People' — groups for which members could not be labeled as 'Person'
- 'Person?' — when clear identification of a pedestrian is ambiguous

In addition to the locations of the individual pedestrians, this dataset provides information regarding the occlusion levels. Therefore, two bounding boxes, b^f and b^v, are assigned to each occluded pedestrian. b^f denotes the full extent of the entire pedestrian, and b^v delineates the visible region in b^f. The fraction of occlusion is then computed as $1 - a_v$, with $a_v = \frac{area(b^v)}{area(b^f)} \in [0, 1]$.

Evaluation methodology. A detection method accepts an image as input and returns a set of bounding boxes ($\mathcal{B}$) and the corresponding scores. The method should perform multiscale detection and any necessary NMS for suppressing or merging nearby bounding boxes, see section 2.5. In the following, we refer to the final detection bounding boxes as $\mathcal{B}^d = \{b_1^d, \ldots, b_{N_B}^d\}$ and the ground truth bounding boxes as $\mathcal{G} = \{b_1^g, \ldots, b_{N_G}^g\}$. Note that $\mathcal{G}$ includes only the fully extended bounding boxes, b_i^f.

A detection bounding box b_i^d and a ground truth bounding box b_j^g form a potential match if they sufficiently overlap. Caltech employs the PASCAL criterion [48] for this evaluation, which states that their area of overlap (a_o) must exceed 50%: $a_o \equiv \text{IoU} = \frac{area(b_i^d \cap b_j^g)}{area(b_i^d \cup b_j^g)} > 0.5$. Each b_j^g can be matched, at most, once, i.e., if an instance, b_j^g, is matched ($a_o > 0.5$) by multiple b_i^ds, the detection bounding box with the highest score is counted as the true positive and the remaining as false positives. Moreover, if any b_i^d matches multiple b_j^gs, the ground truth with the highest a_o counts as the true positive and the remaining count as false negatives. Unmatched b_i^ds count as a false positives and unmatched b_j^gs as a false negatives. In addition to the above, Caltech excludes ambiguous regions from the performance evaluation. These are, for example, instances labeled as 'People' or 'Person?', where the locations of the individuals are unknown, see figure 3.2 (a). To achieve this, the corresponding b_j^gs are marked as *ignore*. The matching criterion for ignored regions is less strict and given as $a_o^{ignore} = \frac{area(b_i^d \cap b_j^g)}{area(b_i^d)} > 0.5$. Ignored b_j^gs do not have to be matched, however, matches are not considered mistakes, i.e., matched b_j^gs do not count as true positives and unmatched b_j^gs do not count as false negatives. Moreover, multiple b_i^ds can match an ignored b_j^g, and a b_i^d can match any subregion of an ignored b_j^g. If a single b_i^d matches multiple b_j^gs, not ignored b_j^gs are preferred. Four types of b_j^gs are always set to ignore: b_j^gs that 1) have heights less than 20 pixels, 2) are truncated by image boundaries, 3) contain a 'Person?', or 4) contain 'People'. For comparing different detection methods, the Caltech benchmark uses a log-log graph indicating the miss rate plotted against the number of false positives per image (FPPI), where miss rate $= \frac{|FN|}{|FN| + |TP|} \in [0, 1]$, with $|FN|$ and $|TP|$ denoting the number of false negatives and true positives, respectively. This curve is created by varying the classifier threshold in the range of $(-\infty, +\infty)$ and considering only b_i^ds with scores greater than the threshold. To summarize the detector performance, the *log-average miss rate* (MR) is computed by averaging the miss rates at nine FPPI rates evenly spaced in log-space in the range 10^{-2} to 10^{0}. For curves that end before reaching a given FPPI rate, the minimum miss rate achieved is used.

Evaluation settings. Caltech offers different settings for reporting the detection performance. The setting described above is referred to as *overall*, see [40]. Other settings use the ignore mechanism to change the evaluation protocol and report results on a selected portion of the test data, such as *near scale* or *occluded*. The portion that is commonly used to rank methods is known as the *reasonable* setting. It evaluates performance on pedestrians with heights over 50 pixels under partial or no occlusion, $a_v \in [0.65, 1]$.

Figure 3.1 (a)-(d) depict the most relevant statistics of the Caltech training set using SI=3 and the reasonable setting. The aspect ratio of the ground truth bounding boxes is given as $\frac{b_w^g}{b_h^g}$, where b_w^g and b_h^g denote the width and height of each bounding box, respectively. (d) displays the expected center location of the pedestrian ground truth bounding boxes and figure 3.2 (a) shows a sample image.

3.2.2   Kitti benchmark

The Kitti dataset [60] became available in 2012. In addition to detection, the dataset offers other benchmarks in the field of computer vision, such as odometry, tracking, and semantic segmentation. Kitti's object detection dataset consists of 7481 training images and 7518 test images, comprising 80256 labeled objects. The images were captured while driving in a German mid-size city, in rural areas, and on highways. The recordings have a resolution of 512×1392.

Annotations. Unlike Caltech, the Kitti detection benchmark considers not only pedestrian but also car and cyclist detection. There are nine types of instances that have corresponding bounding boxes, labeled as:

- 'Car'
- 'Van'
- 'Truck'

- 'Pedestrian'
- 'Person sitting'
- 'Cyclist'

- 'Tram'
- 'Misc'
- 'Don't care'

In this database, 'Don't care' denotes regions where objects have not been labeled. Further, each instance to be detected has two variables that provide the occlusion level: 1) truncation from zero (non-truncated) to one (truncated) refers to an instance leaving image boundaries and 2) occlusion

$\in \{0, 1, 2, 3\}$ referring to {fully visible, partly occluded, largely occluded, unknown}, respectively.

Evaluation methodology. Kitti's matching criteria are similar to those employed in Caltech, except for the following. For all detection tasks, any match to a b_j^g labeled as 'Don't care' does not count as a false positive or true positive. For pedestrian detection, a match to a b_j^g labeled as 'Person sitting' does not count as a false positive or true positive. This is mainly due to the similarity in appearance between these two classes. Contrary to Caltech, 'Cyclist' is considered an independent class and a match to a b_j^g of this class counts as a false positive. Moreover, for all types of b_j^gs, the matching criterion is $a_o > 0.5$, i.e., the evaluation protocol does not employ the ignore regions criterion. The Kitti benchmark compares different methods using a precision-recall curve, where precision $= \frac{|TP|}{|TP|+|FP|} \in [0, 1]$ and recall $= \frac{|TP|}{|TP|+|FN|} \in [0, 1]$, with $|FP|$ denoting the number of false positives. The performance is summarized by computing the *average precision* (AP) defined as the mean precision on a set of eleven equally spaced recall levels.

Evaluation settings. The Kitti benchmark has three evaluation protocols, namely *easy*, *moderate*, and *hard* degrees of difficulty. They differ in the height and occlusion level of the pedestrians considered. The moderate setting is used to rank the methods and includes pedestrians with a height above 25 pixels, a maximum occlusion of 1 (partly occluded), and a maximum truncation of 0.3.

Figure 3.1 (a)-(c) and (e) depict the relevant statistics of the Kitti training set using moderate difficulty. (e) displays the expected center location of the pedestrian ground truth bounding boxes in a lattice of size 376×1242; the image resolutions vary in the range of $[370, 376] \times [1224, 1242]$. Figure 3.2 (b) shows a sample image.

3.3 Traffic sign samples

Vision cameras convert light intensity into an electric signal. To achieve this, they make use of photosensors that have minimal or no sensitivity to wavelength. To capture color information, a mosaic color filter is placed over the pixel sensors. The Bayer filter, for instance, provides information regarding the intensity of the light in the red, green, and blue wavelength regions. In automotive, the color red is frequently a clue to localized regions of interest. Hence, in automotive, a red-clear-clear-clear filter is frequently employed. In this study, we use traffic sign samples captured using such a filter. We discuss these samples in more detail in chapter 6.

Chapter 4

Aggregated Channels Network
A Novel Approach for Detecting Pedestrians in Real Time

4.1 Motivation

Pedestrian detection is a canonical case of object detection with a significant relevance to a number of applications in robotics, surveillance, and advanced driver assistance systems among other fields. Owing to the diversity of the appearance of pedestrians, including clothing, pose, occlusion, and background clutter, pedestrian detection has been considered one of the most challenging tasks of image understanding for decades.

To address the task of pedestrian detection, a wide spectrum of approaches have been used in recent years [11, 40, 61]. As surveyed in [11], existing approaches addressing pedestrian detection can be categorized into three different families: decision forests [8, 34, 122, 196, 210], deformable parts models [49, 50, 129, 134, 201], and deep networks [128, 2, 19, 80, 104, 198, 203]. Within these families, owing to the recent advances in deep learning and parallel computing hardware, deep networks have exhibited impressive accuracy gains compared to the other approaches [21, 74, 94, 168, 177]. In particular, in pedestrian detection, CNNs have monopolized the state of the art. However, to achieve satisfying performance, the commonly used CNN-based pipelines rely on architecturally massive CNNs that are computationally expensive. This shortcoming largely prevents their deployment in commercial applications such as advanced driver assistance systems (ADAS), where the use of a costly GPU is not yet practical. This motivates us to design a CNN-based detector with a considerably smaller network size that can achieve high performance.

47

We propose a competitive arrangement of a fast detector and CNN for pedestrian detection that is both accurate and executes in real time using only a single CPU core. To achieve this, we substantially reduce the total number of multiplications caused by the CNN. In particular, we use the fast ACF detector to generate proposals and introduce a new pipeline that enables the detector to share the ACF planes with the CNN. We use the acronym ACNet (*Aggregated Channels Network*) for the proposed pipeline. To evaluate the compressed proposals collected by the ACF detector, we design a simple and small network, with $\approx 2 \cdot 10^5$ trainable weights, that performs at a low computational cost. The proposed network is trained from scratch, without initializing the weights from pretrained models and without using additional data for the training. Our implementation is based on the open source Computer Vision library [33].

The remainder of this chapter is organized as follows. First, we review the state-of-the-art methods in pedestrian detection and describe briefly our contribution. In section 4.2, we introduce the main components of the pipeline and describe these components in depth. Section 4.3 discusses the datasets used, training procedure, and introduces variants and extensions of the proposed pipeline. In section 4.4, we evaluate all relevant variants of the proposed method and compare the obtained results to the most recent methods in pedestrian detection. Finally, in section 4.5, we provide our conclusions.

4.1.1   Related work

Most recent CNN-based object detection approaches focus on using a two-stage cascade approach. In the first stage, a fast method generates a significant yet reduced number of high-quality proposals that later, in the second stage, are evaluated by the CNN.

In [2, 80, 209, 145, 146], the authors rely on different instances of the well-known channel feature detector family [8, 37, 34, 122, 210] to generate proposals. In all the cases, the best results are reported when using a pretrained CNN model [2, 80, 94], i.e., a large dataset such as Imagenet [31] is used to train the model. The model is afterward adapted and fine-tuned to the target dataset. A similar method is also proposed in [19]. In this case, the authors use a pretrained VGG [168] to examine proposals originating from the first stage. The first stage consists of a boosted decision tree classifier where the number of its multiple channel features is extended to incorporate features originating from a second also pretrained CNN, namely the AlexNet [94]. To combine features with different computational costs in an efficient manner, the authors propose a new scheme to arrange the weak classifiers based on the feature complexity.

Regions with CNN features (R-CNN) methods [67, 68, 144, 104] have recently received considerable attention owing to their success in object recognition tasks. One of the most remarkable examples, Fast R-CNN [67], shares convolutions across proposals, which significantly decreases the computational cost of the network. In this manner, the majority of the computation time is expended to generate the proposals. Ren et al. in [144] extend this idea and design a region proposal network that efficiently shares the features with the Fast R-CNN. Other extensions such as [104] modify the original Fast R-CNN network architecture to incorporate a scale-aware scheme aimed to improve the performance among different scales, i.e., small and large proposals.

In [202, 207], the lower layers from a pretrained network are used to generate features and a boosted forest is built on top of them. It has been demonstrated that after a certain depth of the network, the performance degrades and the CNN features become less discriminatory.

Similar to our idea, a small number of methods in the literature have investigated the use of handcraft features such as HOG [30] or CSS [167] as the CNN input [127, 130, 205], and other color spaces, YUV instead of RGB, [109, 164]. Differing from these approaches, the proposed method is built on top of aggregated channels and the use of a reduced CNN, which efficiently reuses the features computed in the first stage and processes the shrunk input.

4.1.2 Contribution

Within this study, we propose a novel detection pipeline and explore the possibility of reusing aggregated channel features computed in the proposal generation stage. For a fast evaluation of the proposals, features that are created in the first stage as input to the boosted forest classifier, are shared across the entire pipeline. This makes the two-stage cascade more efficient. Candidate patches that pass the first classifier are considered as proposals and forwarded to a small-sized CNN. For this, we design a new CNN architecture that has a sufficient depth to perform accurate pedestrian detection while being able to address the small size of the shared features, i.e., 16×8 pixels for each channel. Moreover, we demonstrate that the pipeline is not restricted to the above size of channel features and can be easily adapted to function with any size of channel features.

We demonstrate that the pipeline has multiple advantages. The input of the CNN is already resized, pre-processed, and compressed by the boosted forest detector. Thus, proposals passing the first stage can be directly inputted to the second stage. Implicitly, the revisiting of the corresponding raw RGB candidates and resizing of large patches is skipped. In addition, the small input size of the CNN dramatically reduces the number of multiplications caused by the convolution layers in such architectures. Consequently, the approach can achieve real-time performance without the use of GPU computation. We introduce multiple variants of the pipeline and demonstrate that investing additional computational effort enables state-of-the-art accuracy. In fact, when compared with the state of the art, even with methods running on a GPU, the proposed method ranks among the leading methods and achieves a highly competitive runtime on a single CPU core.

4.2 Aggregated channel network

The main goal of cascading a fast detector with a strong CNN is to generate, out of the total number of proposals in an image, a significant yet reduced number of high-quality proposals that later can be evaluated by the CNN. The detector collects proposals that likely belong to the target class; these are then rescored by the CNN. The final performance of the arrangement depends highly on the performance of the proposal generation method, see section 2.2.1.5. Modifying the operational point of the detector allows increasing or decreasing the number of proposals, and therefore, the cost caused by the two components. In the following sections, we describe the proposed two-stage cascade design and implementation details.

4.2.1 Proposal generation

For the first stage of our pipeline, we use the well-known ACF detector [34], see section 2.4.0.3. The ACF detector creates a feature pyramid by dividing ten computed channels into 4×4 blocks and aggregating pixel values in each block. These channels are normalized gradient magnitude (1 channel), histograms of oriented gradients (6 channels), and LUV color (3 channels). Every value in these aggregated channels represents a feature value. This allows the detector to decouple the computational cost of the feature creation from the number of decision trees in the cascade, which makes the ACF detector one of the fastest existing detectors. With a detection window size 64×32 and 4×4 aggregation, the ACF detector creates proposals of size $16 \times 8 \times 10$ (1280 features), which is a compression of the input proposal by $\frac{64\times32\times3}{16\times8\times10} = 4.8$ in terms of number of pixels. For the detection, 2048 trees with depth 2 are applied to each detection window in a constant soft cascade manner, see equation 2.4.

4.2.2 Proposal evaluation

Recently, a variety of architectures have been proposed and tested successfully for pedestrian detection, such as CifarNet [93], AlexNet [94] and, more recently, VGG [168]. Although such architectures were not initially addressed for pedestrian detection, they have been adapted and retrained for this specific task. The adaption mainly includes architecturally modifying the top layers. This is performed either to fit the network to the new input size and/or to fit the number of the output neurons to the number of the classes in the target dataset. It is established that the pretraining followed by domain-specific fine-tuning enables faster convergence and commonly superior generalization ability of the network [80, 68, 203, 21]. However, unlike the above networks, the proposed network receives a low-level feature representation as input, which makes it infeasible to adopt such pretrained network architectures for our purposes.

In our pipeline, the CNN evaluates the same windows that the detector examines in the previous stage, i.e., the same candidate windows produced in the feature pyramid by the detector. This is done by sharing the feature pyramid between the detector and CNN and therefore the detector offers candidates with a resolution of $16\times8\times10$. To evaluate the offered proposals, we design a small CNN that we architecturally derive from the well-known CifarNet [93]. We modify CifarNet, which was recently modified and successfully tested for the case of pedestrian detection [80], to fit the predefined input size. The CifarNet, when compared to other networks used for pedestrian detection, is one of the smallest CNNs with $\approx 10^{15}$ trainable weights. Originally designed to solve the CIFAR-10 classification problem [93], it was trained using ten object classes given as color images with a resolution of $32 \times 32 \times 3$ pixels. By studying the approach used in [80], we can immediately observe that the computational cost caused by the max pooling and nonlinearity layers can be ignored when compared to that of the convolution layers. For instance, for a proposal size of $128 \times 64 \times 3$, commonly used for deep architectures [80, 19], CifarNet with minor adaptation to the detection task results in ≈ 100 million multiplications to obtain a classification score. We focus on increasing the speed by decreasing the number of multiplications. An intuitive method to achieve this is either to decrease the number of trainable weights in the CNN or to decrease the dimension of the input image. As mentioned in section 4.2.1, ACF planes compress the input patch. This already enables decreasing the number of multiplications significantly. However, the architecture of the network needs first to be modified in order to deal with such reduced input sizes. To combine the CifarNet with the ACF detector, we change the CifarNet architecture in different aspects which we describe next.

First, inspired by [171], we remove the subsampling layers while retaining the nonlinearities (ReLUs) between the convolution blocks such that the network can still have a sufficient depth and thus remain able to induce abstract representations from the input. In addition to having a positive impact on the learned filters by making them more insensitive to translation, subsampling layers are used to successively reduce the dimension of the input to one dimension at the fully connected (FC) layer, where the output neurons have a receptive field of size equal to the input resolution, allowing the classification task to be performed. By removing them, dimension reduction can be performed in a more gentle manner through the convolution filters $f : \mathbb{R}^{H \times W \times C} \rightarrow \mathbb{R}^{H' \times W' \times C'}$, where $H^{(\prime)}, W^{(\prime)}$, and $C^{(\prime)}$ refer to the height, width, and depth of the input and output. Given N_f filters of dimension $H'' \times W'' \times C''$, without the use of zero-padding and with strides $\delta^w = \delta^h = 1$, the convolution operation results in an output volume of dimension $H' = H - H'' + 1, W' = W - W'' + 1$, and $C' = N_f$.

Secondly, to avoid the disproportionate use of zero-padding and retain reduced computational time, we change the filter sizes in CifarNet from $C_1 = C_2 = C_3 = 5 \times 5$ to 5×3, where C_i denotes the size of the kernels in the ith convolution layer.

Next, we replace the regularization layers (i.e., contrast normalization, see [80]) with a computationally less expensive dropout layer [172] placed at the top of the network.

Finally, we replace the nonlinearity of the FC layers by sigmoid instead of ReLU and change the objective function of the CNN from softmax and cross entropy to the mean squared error, see section 2.2.2.4. The reason for the last change is because we observed that the sigmoid nonlinearity combined with the new loss function led to a smoother learning behavior and a marginally improved performance in our initial tests.

Figure 4.1 illustrates the architecture of the ACNet in detail. The input, output, and kernel volumes in figure 4.1 reflect the proportionalities in a realistic manner. As can be observed, the total computational cost of the network is a fraction of that of CifarNet with a RGB input patch of size $128 \times 64 \times 3$.

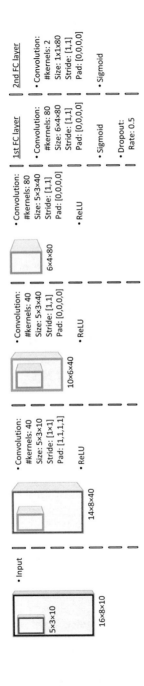

Figure 4.1: ACNet architecture with a total number of multiplications in the convolution layers of $\approx 3 \cdot 10^6$. The input, output, and kernel volumes reflect the proportionalities in a realistic manner. #kernel= N_f and size:(H'', W'', C'') denote the number and dimension of the filters in each convolution layer, where the dimension is given by height, width, and depth, respectively. The parameter stride: $[\delta^w, \delta^h]$ defines the strides of the convolution kernel in the horizontal (δ^w) and vertical (δ^h) directions and the parameter pad=[top, bottom, left, right] denotes the number of zeros used to extend the input dimension before the convolution is computed. Dropout with rate:0.5 means that 50 percent of the connections are dropped out.

4.3 Implementation details

4.3.1 Dataset

Dataset. It is well known that for CNNs the number of training samples is reasonably important to achieve acceptable performance, especially when they are trained from scratch. In our experiments, we use two of the largest and most challenging datasets publicly available in the literature, namely the Caltech [39] and Kitti [60] datasets, discussed in section 3.2.

Validation. In the Caltech dataset, the training data consist of six video sequences. We follow a 4-fold cross-validation strategy and divide the training set into four subsets. This is performed in a manner that the number of pedestrians considered in the reasonable setting (see section 3.2.1) is approximately the same in each subset. In total, four detectors are evaluated using three subsets for training and the remaining set for validation. Henceforth, we refer to those sets as validation training set and validation test set.

In the Kitti dataset, we divide the training set into four subsets in such a manner that the number of pedestrians considered in moderate difficulty (see section 3.2.2) is approximately the same in each subset. We follow a 4-fold cross-validation strategy, train, and test four detectors using three subsets for training and the remaining set for testing.

4.3.2 Operational point

It is possible to adopt the ACNet architecture into the process chain before and after NMS. Furthermore, the boosted forest classifier can be calibrated to achieve a desired true positive rate. The lower the cascade threshold θ^* is, the more samples pass the cascade and consequently, the higher true and false positive rates will be, see equation 2.4. Because the proposed CNN applies directly after the detector and evaluates only samples that pass the detector, its performance depends closely on that of the detector. For instance, when the detector, at a chosen operational point, achieves a true positive rate of 90%, this rate becomes the upper boundary of the entire pipeline and the CNN can only move the operational point (at that true positive rate) at best to a lower false positive rate.

On Caltech, we train our baseline detector according to [34]. For the Kitti dataset, we use the same setting as for Caltech, but add one octave at the top of the feature pyramid. This is because the pedestrians to detect in Kitti have heights above 25 pixels, in Caltech above 50 pixels, see section 3.2. Adding one octave at the top of the feature pyramid allows for including these pedestrians using the same template size. This causes the input image to be first upsampled by 2.

In figure 4.2 (a), we apply our baseline detector on the Caltech validation test set before and after NMS. Ideally, we seek an operational point where the detector returns a high number of true positives and a low number of false positives. As can be observed, applying NMS decreases both true and false positive rates noticeably. To obtain the same true positive rate before and after NMS, we could reduce θ^* in the case of "after NMS". This would cause a greater number of candidates to pass the entire chain of the binary trees. Thus, the detector runtime would decrease.

4.3.3 Collecting training set

We train ACNet with different training sets by varying the IoU metric(see section 3.2), which measures the matching quality of the detections. We collect two negative training sets using the false positives of our baseline detector with an IoU below 0.5 and below 0.3. Because the proposed AC-Net architecture does not have any subsampling layer, it could behave sensitively to translation. To match the translation in the positive candidates that are offered by the detector, we use the detector to collect positive jittered samples from the training set. For every ground truth, we accept the detection with the highest score and an IoU > 0.5.

Tables 4.1 and 4.2 display the results on the Caltech and Kitti validation test sets, respectively. In both experiments, the ACNet architecture is directly applied after the ACF detector, before and after NMS. In both datasets, we notice that using negative samples with more overlap with ground truths and positive jittered samples appears to be beneficial.

By studying figure 4.2 (a), we identify that ACNet benefits from a high true positive rate. This is also indicated in tables 4.1 and 4.2. Figure 4.2 (b) shows that in the case of "before NMS", the ACNet architecture receives more positive candidates around each pedestrian. Figure 4.2 (c) shows the results of an experiment which we perform on the test set in order to understand if the division into positive and negative training sets using 0.5 as the IoU threshold can be a harsh criterion. This experiment is discussed in more depth in section 4.4.

We attempt different operational points for our baseline detector by recalibrating the cascade detector and conclude that restricting the number of candidates to an average of 40 per image provides a sufficient true positive

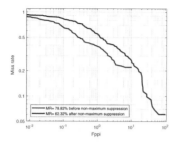
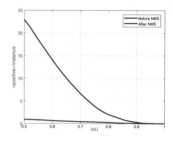

(a) Baseline detector before and after NMS on Caltech validation test set.

(b) Averaged number of positive proposals (proposals with IoU >0.5) per pedestrian on Caltech validation test set, before and after NMS.

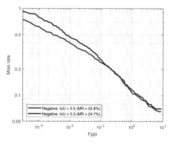

(c) Evaluation of two fully trained AC-Net+ pipelines. Both are trained on Caltech training set and evaluated on its test set using the reasonable setting.

Figure 4.2: (a) and (b) depict the behavior of the baseline detector before and after NMS. (c) displays the impact of the negative training set on fully trained pipeline.

rate and number of positive proposals per pedestrian. More concretely, the ACF detector achieves, at this operational point, a true positive rate of more than 92% on the validation test set, see figure 4.2 (a). We use these settings for all experiments in the following sections.

We repeat the above experiments using two common pre-processing steps [27]: 1) we normalize the CNN training data by subtracting the mean and 2) we divide the centered samples by the standard deviation of the training samples. The last scheme corresponds to local brightness and contrast normalization. Although such normalization methods are usually beneficial for shallow networks, in both scenarios we observe no significant impact on the performance. In favor of a faster runtime, we do not investigate if this

Positives	Negatives	MR[%] (BN)	MR[%] (AN)
$\mathcal{G}$	IoU< 0.5	41.9	50.1
$\mathcal{G}$	IoU< 0.3	45.0	51.7
$\mathcal{G}$, IoU > 0.5	IoU< 0.5	*41.4*	46.9
$\mathcal{G}$, IoU > 0.5	IoU< 0.3	42.7	47.7

Table 4.1: Performance of ACNet on Caltech validation test set trained using different training sets. MR: log-average miss rate (lower is better). BN: ACNet applied before NMS. AN: ACNet applied after NMS. $\mathcal{G}$: ground truth bounding boxes. IoU: intersection over union of detections and ground truths bounding boxes.

Positives	Negatives	AP[%] (BN)	AP[%] (AN)
$\mathcal{G}$	IoU< 0.5	70.4	65.9
$\mathcal{G}$	IoU< 0.3	69.2	65.1
$\mathcal{G}$, IoU > 0.5	IoU< 0.5	*75.6*	68.9
$\mathcal{G}$, IoU > 0.5	IoU< 0.3	68.0	67.2

Table 4.2: Performance of ACNet on Kitti validation test set using different training sets. AP: average precision (higher is better). BN: ACNet applied before NMS. AN: ACNet applied after NMS. $\mathcal{G}$: ground truth bounding boxes. IoU: intersection over union of detections and ground truths bounding boxes.

changes when involving further hard negative mining iterations.

4.3.4 Training ACNet

For training ACNet, we follow the bootstrapping strategy. In stage 0, we collect false positives with IoU < 0.5 and jittered positives with IoU > 0.5 from the training data using the detector. These samples plus the ground truths constitute the initial training set for the ACNet architecture. We train the architecture fully, adapt it to the pipeline, start the first stage of bootstrapping using ACNet to collect more hard negatives and continue training the network with the additionally collected samples. We continue bootstrapping until the network saturates and performance does not improve.

Even using Caltech10×, pedestrians are rare in the dataset. The number of the collected negative samples can rapidly rise to a multiple of that of the positive samples. To address this imbalance, we restrict the number of collected negative samples by setting at FPPI (false positive per image) = 1 a threshold for the ACNet architecture. This threshold is calculated before entering each bootstrapping stage and applied for its entire duration. This ensures that only one false positive per image on average is collected in each stage.

The network is randomly initialized with Gaussian with mean zero and standard deviation 0.2 and trained from scratch. We use stochastic gradient descent with a mini-batch size of 1000, a weight decay of 0.0005, and a momentum of 0.9. The learning rate started from 0.02 and is divided by 10 when the error plateaus.

4.3.5   Extending the pipeline

The proposed pipeline can be adapted to any proposal generator method that creates channel features [8, 37, 34, 122, 210] and can itself be efficiently used as such a method. In the following, we first modify the pipeline to allow the adoption of proposals of a different size and demonstrate three extensions of the existing pipeline.

ACNet+. To study the limits of a stronger proposal generator, we explore the use of the ACF+ detector [122]. Compared to ACF, the ACF+ detector uses more data for the training, 4096 depth 5 trees for the binary classification, and 2×2 aggregated channel features. The detection window size is as before, 64×32, which creates candidates of size $32 \times 16 \times 10$ and results in 5120 features per candidate. Because of the different aggregation parameters, it is not possible to use the trained ACNet architecture after the ACF+ detector. To combine the ACNet architecture with the ACF+ detector, we change the padding of the first convolution layer and append a subsampling layer directly after it. In this manner, all following layers and their computational cost remain as described in section 4.2.2. The ACNet+ architecture causes $\approx 5 \cdot 10^6$ multiplications in the convolution layers. We call the new pipeline *ACNet+*.

We train ACNet+ following the same procedure as described in section 4.3.4. Using ground truths, jittered positive samples with IoU > 0.5, and negative samples with IoU < 0.5 provide the best performance.

ACNetBF/ACNet+BF. Inspired by [207, 202], we use convolution responses as features for the boosted random forest classifier. Unlike the mentioned works, these features are not induced from a RGB input; rather, they are low-level channel feature representations. We enrich the detector proposals with additional convolution features acquired from the corresponding network. We do this by vectorizing and concatenating the network input

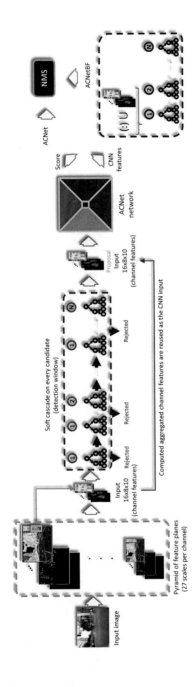

Figure 4.3: ACNet and ACNetBF pipelines. The pipeline can be divided into three processing steps. First, the feature pyramid, created by the detector to detect at different scales (orange dashed box). Next, the constant soft cascade that every content of the detection window must pass (in blue dashed box). In the cascade phase, when the score of the content falls below a predefined threshold, the detection window is rejected. Finally, the proposed ACNet (top right) and ACNetBF (bottom right in violet dashed box) architecture, where every detection window that passes the soft cascade is considered a proposal and, therefore, evaluated to obtain a new score.

and all three intermediate convolution outputs, see figure 4.1, to create a feature vector $\mathcal{F}$. In the case of ACNet, we create $\mathcal{F}$ of length $|\mathcal{F}| = 10080$ by first vectorizing $16 \times 8 \times 10$, $14 \times 8 \times 40$, $10 \times 6 \times 40$, and $6 \times 4 \times 80$ features from the input, first, second, and third convolution layers, respectively, and then concatenating them. In the case of ACNet+, we use the response of the first convolution layer after the pooling layer to ensure that only the size of the detector's proposal ($32 \times 16 \times 10$) differs. This results in $|\mathcal{F}| = 13920$. The vector $\mathcal{F}$ is input to a boosted forest (BF) consisting of a maximum 4096 decision trees with a maximum depth of 5 — maximum indicates that we prune the depth of the trees and length of the cascade if the error falls below a given threshold. The boosted forest trains via discrete AdaBoost 2.2 and forms a monolithic cascade 2.4. Proposals passing the first stage (our baseline detector) are forwarded into the network and are rescored by the BF. This BF also applies before NMS and does not reject samples at any point $\Leftrightarrow \theta^* = -\infty$. We refer to these extensions as *ACNetBF/ACNet+BF*. Both pipelines, ACNet and ACNetBF, are depicted in figure 4.3. We train the BF using the same bootstrapped training set that we use for training the underlying ACNet/ACNet+ architecture and do not further bootstrap. Furthermore, a random subset of the size $|\mathcal{F}| \cdot 1/16$ is used to create every non-leaf node in each tree.

ACNet+BFResNet. In [209], it is demonstrated that the performance of deep CNNs depends closely on the quality of the proposal generator. Considering the effectiveness of ACNet and its variants, these can be used as proposal generators. To validate this point, we extend the pipeline using a ResNet50 model [74] pretrained on ImageNet [31]. ResNet50 accepts cropped RGB image patches of size $224 \times 224 \times 3$. However, such large patches are computationally overly expensive and not suitable for pedestrian detection [80]. We modify the network by removing the three top layers (FC, average pooling, and ReLU layers) coming directly after the last sum layer. We then randomly initialize two FC layers with 512 and 2 units, respectively, and fine-tune the network using [186]. This allows us to use the canonical $128 \times 64 \times 3$ size for the pedestrian template [80]. In this window, the pedestrian occupies an area of 96×48. We attach the ResNet50 to the ACNet+BF pipeline after NMS, where the number of proposals is further significantly reduced through the NMS algorithm 2.5. We refer to this pipeline as *ACNet+BFResNet*.

For training ACNet+BFResNet, we use a similar strategy as for ACNet 4.3.4. We employ bootstrapping and start training ResNet50 at stage 0 using samples that we collect with ACNet+BF. As before, we use a threshold at FPPI $= 1$ for collecting. The initial set consists of ground truths, jittered positives with IoU > 0.5, and false positives with IoU < 0.5. We train ResNet50 by first training the two new FC layers until the error plateaus and then training all layers together. After being fully trained, we use AC-Net+BFResNet, start the first stage of the bootstrapping to collect more hard negatives and continue training the ResNet50 model. We apply bootstrapping until no further improvement is possible. We train the model itself using the parameters suggested by the authors [74], in addition to using a lower learning rate. We use stochastic gradient descent with a mini-batch size of 256, a weight decay of 0.0001, and a momentum of 0.9. The learning rate starts from 0.01 and is divided by 10 when the error plateaus.

4.3.6 Improving the quality of the proposals

4.3.6.1 Bounding box regression

To study the quality of the proposals provided by the detector, we define three parameters

$$\rho_x = \frac{b_x^g - b_x^d}{b_h^d}, \quad \rho_y = \frac{b_y^g - b_y^d}{b_h^d}, \quad \text{and} \quad \rho_s = \frac{b_h^g}{b_h^d}, \tag{4.1}$$

where b_x^d, b_y^d are the coordinates of the top-left corner of the detection bounding box and b_h^d is its height on the RGB image. We normalize these parameters to be in the range $[0, 1]$ and denote the normalized parameters as $\overline{\rho}_x, \overline{\rho}_y$, and $\overline{\rho}_s$.

Figure 4.4 depicts the matching quality of the baseline detector using the three normalized parameters. As can be observed, the detector appears to have a characteristic error. The location and scale deviations of the detection bounding boxes are not symmetrically distributed. The reason for this behavior could be related to the discrete manner of how the detector samples proposals, i.e., the sliding window approach. It could also be related to or strengthened through bad annotations of the pedestrians [209].

In the following, we explore the possibility of improving the quality of the proposals that have passed the detector by applying bounding box regression on them before being forwarded to the ACNet architecture.

As discussed in section 4.2.2, the ACNet architecture has a loss function that allows performing regression tasks. Unlike traditional approaches, we perform the regression directly on the ACF channels using the labels that point to the corresponding RGB patches, namely $\overline{\rho}_x$, $\overline{\rho}_y$, and $\overline{\rho}_s$. In the testing phase, we first estimate the new position of each proposal then translate this position to a corresponding patch on the feature pyramid. However, this practice includes limitations that we must consider. The estimation of $\overline{\rho}_s$ must be rounded to the nearest scale in the feature pyramid. Further,

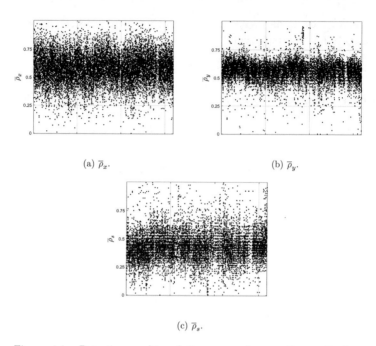

(a) $\overline{p}_x$.

(b) $\overline{p}_y$.

(c) $\overline{p}_s$.

Figure 4.4: Detection quality of the proposal generation method on Caltech10× training set.

the estimation of $\overline{p}_x, \overline{p}_y$ must be a multiple of the channel aggregation value, see section 2.4.0.3. To overcome these limitations and first determine the extent that we could improve through the use of such a method, we use $\overline{p}_x$, $\overline{p}_y$, and $\overline{p}_s$ without the above limitations. We, therefore, create channel features for each candidate using the predicted new RGB location. This step is costly and repeats with the number of candidates in an image. Nonetheless, it allows exploring the upper-performance boundary.

We develop two different approaches for the regression task. For the first, we use a regression network by customizing the pretrained ACNet architecture. Therefore, we remove the last FC layer and initialize a new one with three output neurons for predicting the parameters $\overline{p}_x, \overline{p}_y$, and $\overline{p}_s$. In the second approach, the regression is performed using a SVM. For this, the SVM receives the last convolution outcome of the modified ACNet as input.

Training occurs in a similar manner as for the ResNet50 model. We first train the new initialized layer and then fine-tune the entire network. We consider two different training sets. We collect the first set, $\{(x^{(i)}, y^{(i)})\}_{i \leq N}$

	Collected	e/n
ρ_x	[-0.16, 0.14]	[-0.15, 0.15]
ρ_y	[-0.40, 0.31]	[-0.40, 0.40]
ρ_s	[0.71, 1.41]	[0.60, 1.40]
E^*	0.0230	0.0242

Table 4.3: Regression on aggregated channel features. The first row refers to the method used to create the training set. "Collected" refers to samples collected using the baseline detector and "e/n" refers to evenly/normally created random labels. ρ_x, ρ_y, and ρ_s are the ranges of the labels and E^* is the lowest achieved error using our regression models.

with $y^{(i)} = (\overline{\rho}_x^{(i)}, \overline{\rho}_y^{(i)}, \overline{\rho}_s^{(i)})$, using the baseline detector. The second set is created via artificially jittered ground truths which are obtained using evenly or normally distributed random labels $y^{(i)} = (\overline{\rho}_x^{(i)}, \overline{\rho}_y^{(i)}, \overline{\rho}_s^{(i)})$. The normal distribution is used to emulate the statistic of the data collected by the detector. Furthermore, we train different regression networks varying the ranges of the unnormalized labels (ρ_x, ρ_y, and ρ_s). This is accomplished by step-wise reducing their initial selected ranges, shown in table 4.3, which is equivalent to allowing less correction in position and scale. The unnormalized labels are applied to crop the corresponding RGB image patches and create their channel features. These channel features are the input, x, to our regression network. By evaluating the IoU > 0.5 criterion for positive samples, the random labels are corrected, if required. Furthermore, we consider differently jittered ground truths for each pedestrian to enlarge the number of training samples.

Table 4.3 gives a summary of the used ranges of ρ_x, ρ_y, and ρ_s for the case of collecting the training samples using the baseline detector ("Collected") and the initial ranges for the case of artificially jittered ground truths ("e/n"). The last row of the table reports the best-achieved error E^* on an independent portion of the training data. In both scenarios, this was achieved using the CNN approach. To identify the meaning of this experiment in terms of performance, we use the regression network before NMS and forward the corrected proposals to the ACNet architecture. We measure an improvement of approximately 1.5 percent points (pp) in MR for the baseline detector. However, the impact on ACNet was marginal, < 0.5 pp in MR. As discussed above, this result is the upper-performance boundary, without considering the previously mentioned limitations. Hence, in favor of a faster runtime, we discard this approach and attempt another which we describe next.

4.3.6.2   Geometrical plausibility of the proposals

Motivated by [112, 134], we remove the bounding boxes, derived from the first stage, that do not have a plausible relationship between their position and scale. To accomplish this, we make use of the same assumptions and ground truths available in the training data. We apply a linear regression to learn the relationship $b_h^g = x_0 + x_1 \cdot (b_y^g + b_h^g)$, where $b_y^g + b_h^g$ is the bottom vertical position of the ground truth bounding box. We then set four parameters, namely h^-, h^+, y^-, and y^+. Using these parameters, we examine the geometrical plausibility by requiring every proposal, b^d, to fulfill

$$y^- < b_y^d < y^+, \tag{4.2}$$
$$h^- < b_h^d - \left[x_0 + x_1 \cdot (b_y^d + b_h^d) \right] < h^+. \tag{4.3}$$

A proposal that violates either of these two criteria is discarded. Using a grid search, we optimize these parameters on an independent portion of the training data. The benefit of this approach is twofold. It removes hard negatives that are located in unusual locations in the image and reduces the number of proposals that must pass the ACNet architecture and NMS. In Kitti, we are not able to use this approach due to the inconsistency of the image resolutions — height and width of the images fluctuate within $[370, 376]$ and $[1224, 1242]$, respectively, see section 3.2.2.

4.4   Experiments

As mentioned in section 4.3.1, we use the enlarged Caltech training set, known as Caltech10×, for the following experiments. We present our results on the Caltech dataset with and without involving the bounding box purging strategy from section 4.3.6.2. We use the suffix "\rmBbs" to highlight the latter case. Note that this strategy is not involved in any of our results on Kitti.

ACNet/ACNet+. On Caltech, we use both baseline detectors according to [34] and [122]. For Kitti, we use the same settings, but we upsample the input image to detect pedestrians shorter than 50 pixels in height, see section 4.3.2. We train both pipelines following the routine described in section 4.3.4. Both variants saturated within 3-5 stages of bootstrapping on both datasets. Figure 4.6 displays that both pipelines outperform their underlying detectors by a significant margin.

At this point, we revisit the open question from section 4.3.3. We completely repeat the training procedure of ACNet+, maintaining all setting as before,

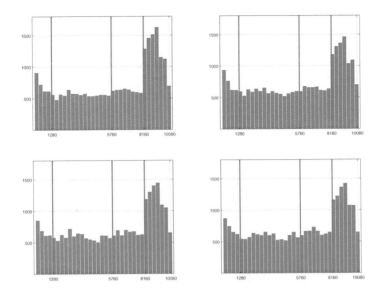

Figure 4.5: Selected feature set used for training ACNetBF on Kitti training dataset. The feature pool consists of ACNet channels and ACF planes. The three vertical lines separate the respective feature types. The features are concatenated in the following order from left to right: ACF planes, and channels from convolution layers 1, 2, and 3. The CNN channels are not aggregated or further processed. From left to right: each histogram indicates the feature distribution of the first, second, third, and fourth 1024 weak classifiers.

while exchanging the negative set and using samples with IoU < 0.3. We do this to determine the effect of this threshold on the performance. Figure 4.2 (c) depicts the outcome of this experiment. We observe that using negative samples with an IoU less than 0.5 leads to lower localization errors at lower FPPI ranges. We know that at higher FPPI ranges, the ability to discriminate foreground and background has an important role [209]. Although the division into positive and negative samples using 0.5 as the IoU threshold appears to be a harsh criterion, it appears to have no negative effect in the discriminating capability of the learned filters at higher FPPI ranges.

ACNetBF/ACNet+BF. As described in section 4.3.5, we use the same bootstrapped training set that was collected during the ACNet/ACNet+ training to train their BF extensions. This ensures that the contribution comes only from the new attached last component. Figure 4.6 indicates that when combining ACF channel features with the CNN channels, the BF has a superior classification ability than the 2 FC layers. Both datasets demonstrate improvement through the extension with virtually no additional computation effort, see figure 4.7. Figure 4.5 depicts the feature selection of the BF extension. It can be observed that the features deriving from the first and second CNN layers are chosen with an almost equal frequency. Their discriminating ability does not outperform that from the low-level ACF representation. Features deriving from the third CNN layer, conversely, which have the largest receptive field and a higher degree of abstraction, appear to be preferred. In particular, in the beginning phase of the cascade, these features appear to be effective in distinguishing the classes and dominating the feature selection.

ACNet+BFResNet. To saturate the ResNet50 model, we create a new negative set and use bootstrapping multiple times, as described in section 4.3.5. Because the ResNet50 model receives RGB patches, it allows the application of common data augmentation techniques. Therefore, we perform different training routines. In the first variant, we strictly repeat the training routine for ACNet. We iteratively collect more negative samples and fine-tune the model. In the second variant, we apply flipping, scaling, and cropping of the positive samples during the mini-batch training. This, we perform randomly with a probability of 50%. Lastly, we repeat the second variant while adding, after each bootstrapping stage, replicas of the ground truths to the training set. The reason behind this action is based on the observation that the deep model requires significantly more bootstrapping stages to fully saturate [19]. Retaining all bootstrapped samples, as we did in the first variant, results in an imbalanced training set at higher bootstrapping stages. We bootstrap 10 times on Caltech. Using the first and second variant, described above, we attain a performance of approximately 17% in MR, the last variant demonstrates an improvement of ≈ 3 pp. On Kitti, we report our results using the last training routine. The performance on both test sets can be seen in figure 4.6.

4.4.1   Comparison to the state-of-the-art methods

Not all methods report their results on both datasets. Figure 4.6 compares the most recent and relevant state-of-the-art pedestrian detection methods. Next, we compare our results in more aspects and with a special focus on the trade-off between runtime and performance. Figure 4.7 displays the methods in an accuracy-speed space.

Two of the recent methods that ACNetBF/ACNet+BF is conceptually most comparable to are *AlexNet Caltech10×* [80] and *DeepCascadeED* [2]. Both

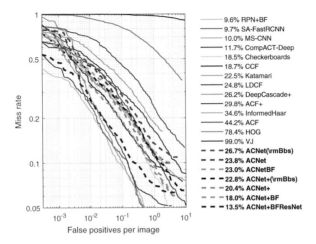

(a) Comparison to state-of-the-art methods on Caltech test set, using reasonable setting. Methods are compared using the log-average miss rate metric (MR) measured in the range $[10^{-2}, 10^{0}]$ false positives per image, lower is better. (\rmBbs) refers to the exclusion of the purging method, see section 4.3.6.2.

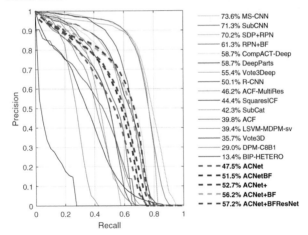

(b) Comparison to state-of-the-art methods on Kitti test set, using moderate difficulty. Methods are compared using the average precision metric (AP), higher is better.

Figure 4.6: Comparison to state-of-the-art methods. Our results are shown in dashed lines labeled bold.

use a fast detector to collect proposals and a CNN to rescore them.

DeepCascadeED creates a three-component pipeline. The first is a boosted forest detector [9, 8] that executes at more than 100 frames per second (FPS) and is known as *VeryFast*, see Figure 4.7 (a). After applying VeryFast, the remaining proposals must pass a small CNN with three hidden layers that further reduce their number. Finally, the proposals are rescored by a deep CNN. This last CNN is reduced version of the AlexNet [94] and pretrained on ImageNet [31]. The authors report that the pretraining improves the performance by ≈ 3 pp. The entire pipeline, including the boosted forest detector, executes on a GPU. Furthermore, additional pedestrian datasets, ETH [47] and Daimler [45], complement the Caltech training set. According to the reported results, this last step permits ≈ 5 pp better performance. Using this setting, DeepCascadeED achieves 26.2% in MR while running at 15 FPS. In comparison to this, ACNetBF executes virtually double that speed on a single CPU core, with ≈ 28 FPS, while being ≈ 3 pp superior in MR.

AlexNet Caltech10x uses *SquaresChnFtrs* [10, 11], that is related to our baseline detector, as the first component. The AlexNet structure is used as the second component to rescore the collected proposals. The structure is trained from scratch. On Caltech, the performance and runtime are reported as 27.5% in MR and ≈ 0.43 FPS, respectively. When using a pretrained AlexNet instead of training from scratch, the authors report improving to 23.3% in MR. Both results are depicted in figure 4.7 (a).

Our ResNet50 extension is conceptually most comparable to the recent methods *CompACT-Deep* [19] and *SP++VGG-VD16* [146].

SP++VGG-VD16 experiments with different proposal generator methods. Moreover, it introduces a more general methodology compared to AlexNet Caltech10×. Instead of using the RGB proposals directly, it considers, in addition to RGB, using different segmentation maps such as gradients, YUV, and LUV. The authors employ a VGG [168] model pretrained on ImageNet [31] as the second component. By evaluating different setups, they conclude that using their strongest detector, *SpatialPooling+* [132] with 21.9% in MR, and RGB patches provide the best result, which is reported to be 16.66% in MR at 0.37 FPS on Caltech. It should be stated that the mentioned time refers solely to the computation time of the last component. However, we use the reported time to compare with our results, see figure 4.7 (a). Our ResNet50 extension executes over 13 times faster and attains greater than 3 pp better MR. Interestingly, the authors report a drop in performance to 23.34% in MR when using the ACF+ detector, our baseline detector, as the first component.

CompACT-Deep reports results on the Caltech and Kitti datasets. The first component of CompACT-Deep is a cascade of decision trees that has access to a large and diverse feature pool containing features with different complexities. To arrange weak classifiers in an efficient manner, the authors propose an algorithm for creating a cascade that considers the feature computation time. The second component of the pipeline is a VGG [168] model pretrained on ImageNet [31]. The VGG includes the pipeline in three different fashions. In the first variant, the VGG applies after NMS, functioning in the same manner as in SP++VGG-VD16. In the second and third variants, the VGG is integrated into the cascade architecture, i.e., into the first component. More precisely, it is considered the last weak classifier in the decision tree chain. Thus, its score adds to the accumulated score from the previous weak classifier. The second and third variants differ only in that the VGG applies once after and once before NMS. The results indicate that when applied after NMS, integrating the VGG improves by ≈ 1 pp in MR on Caltech. Furthermore, when the VGG is integrated, applying it before NMS improves by ≈ 2 pp in MR. CompACT-Deep has a strong proposal generation method, CompACT with 18.92% in MR, and executes its CNNs entirely on a GPU — its feature pool incorporates channels from an AlexNet [94] pretrained on ImageNet [31]. Figure 4.7 indicates that CompACT-Deep performs marginally better on both datasets compared to ACNet+BFResNet. In terms of runtime, ACNet+BFResNet executes on a single CPU twice as fast as CompACT-Deep on a GPU.

Figure 4.7 (a) demonstrates that ACNet+BFResNet is among the four top-performing methods on Caltech. Figure 4.7 (b) displays three of them in virtually the same constellation, except for the method called MS-CNN [18], which leads the Kitti benchmark. This method is specially designed for robust multiscale object detection, in particular for smaller instances. A closer look at the false positives of the proposed method reveals that the majority of these arise at smaller scales. A more careful parameter tuning of the detector and incorporating scale-aware techniques to the CNNs could improve our results.

4.4.2 Runtime analysis

All experiments were conducted in serial mode on a desktop PC with an i5-4690 CPU. We calibrate our baseline detectors as described in section 4.3.3. The calibration has an influence on the runtimes of the detectors and their subsequent components. The new runtimes of the baseline detectors are displayed in figure 4.7. In table 4.4, we decompose the runtimes of all components using the Caltech and Kitti test sets. The imagery on Caltech

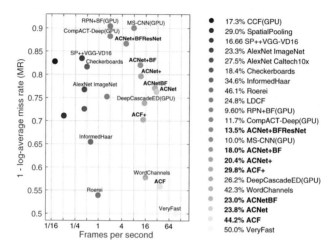

(a) Caltech test set (reasonable setting).

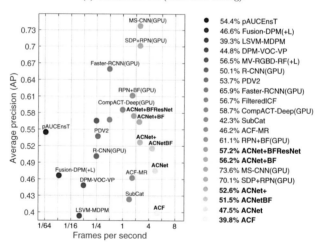

(b) Kitti test set (moderate difficulty).

Figure 4.7: Performance of pedestrian detection methods in accuracy-speed space. Our results are labeled bold. The higher the methods are located, the better the performance. Fast methods are located in the right sphere. In the legend, the methods are ranked after their runtimes — from slow to fast execution time.

has a resolution of 480×640 pixels and pedestrians with heights above 50 pixels. The imagery on Kitti has a resolution fluctuating around 375×1242 pixels and pedestrians with heights greater than 25 pixels. Because of the smaller heights on Kitti, the input image must be upsampled, see section 4.3.2. Owing to the upsampling and higher resolution, methods execute more slowly on the Kitti set. Table 4.4 and figure 4.7 emphasize that the performance gain through the reduced CNN has a low computational cost. This can be further enhanced using a GPU.

Dataset	Method						
	ACF	ACNet	ACNetBF	ACF+	ACNet+	ACNet+BF	ACNet+BFResNet
Caltech	0.026	0.033	0.036	0.070	0.083	0.085	0.196
Kitti	0.156	0.177	0.208	0.371	0.398	0.413	0.555

Table 4.4: Runtime comparison between ACF, ACF+, ACNet, and its extensions on Caltech and Kitti test sets. Times are in seconds per frame obtained using a single CPU core.

4.5 Epilogue

We presented a novel approach that efficiently combines a fast sliding window detector with a newly designed and small-sized CNN. In particular, the pipeline requires performing the heavy feature computation only once and shares features between the detector and the CNN. In this approach, the CNN receives a low-level feature representation as input that has been resized, pre-processed, and compressed by the boosted forest detector. Thus, revisiting of the corresponding RGB candidates is obsolete. Moreover, owing to the compressed input dimension and small CNN size, the number of multiplications caused by the convolution layers in these structures is dramatically reduced. Consequently, the approach can achieve real-time performance without using GPU computation. The simple design and small model size of the proposed CNN allow it to be deployed in low-consumption hardware such as embedded chips. The proposed pipeline has the potential to evolve in different aspects. We would expect to benefit from employing, e.g., an enhanced CNN optimizer, enhanced architectural design, scale- and context-aware mechanisms. In addition, we introduced multiple variants of the pipeline and demonstrated that investing additional computational effort enables state-of-the-art performance. In terms of accuracy, the proposed method ranks among the leading methods for pedestrian detection. In terms of runtime, it competes successfully with methods running on GPUs while running on a single CPU.

Chapter 5

Insatiate Cascaded Boosted Forest
A Novel Approach for Exploiting the Asymmetry of the Dataset

5.1 Motivation

Boosted forest classifiers in a cascade formation have been widely used for object detection such as detection of faces [190, 13, 200, 199], pedestrians [34, 156, 112], and vehicles [157, 6]. To detect a target object at any scale and location in an image, cascaded boosted forest (CBF) detectors typically involve the sliding window paradigm combined with a multiscale image/feature representation. A CBF binary classifier uses a series of binary decision trees, which are created and concatenated by a boosting algorithm, to examine a large number of patches for each given image. Each patch is classified as either positive (matched with a target object) or negative (mismatched with a target object). In this process, a cascade architecture is adopted such that the majority of negative image patches can be rejected by evaluating only a few trees. Their economical consumption of hardware resources makes CBF detectors attractive for many real-world applications such as pedestrian detection for an ADAS [185, 28, 166], which requires high accuracy and more importantly, real-time capability.

Despite many successes in object detection, training a CBF detector with a high performance is not a trivial task. Specifically, the number of hyperparameters in the CBF detector training is large and there is no clear guideline regarding how to set them. Furthermore, the kind and number of training samples required for CBF classifiers is also an unclear point. As is well known in deep learning, a large and diverse training set is one of the key points to achieve an acceptable performance. This is why the majority of deep neural networks are first pretrained on large datasets, such as ImageNet [94], and then adapted to a specific task [80, 206, 104]. This workaround, through transfer learning, enables deep neural networks to achieve excellent performance even in cases where the training data is scarce. However, to the best of our knowledge, there is no comprehensive discussion regarding the quality and quantity of the training samples need to be collected in order to saturate a CBF object detector.

Considering a multiscale detection task, the dataset is given in the form of a finite number of images and ground truths. Figure 5.1 depicts a sample image from the Kitti pedestrian dataset. Pedestrians to be detected are surrounded by a red bounding box. Violet bounding boxes indicate instances from other labeled classes. Using the sliding window paradigm and an image pyramid, each image must be evaluated on more than 600000 different locations. In theory, each location that does not contain a pedestrian can serve as a negative training sample. In practice, only a subset of the negative samples is used for training. This high asymmetry of the dataset is rarely addressed when training a CBF detector. It has been established that the quality of the negative subset has a high impact on the performance, which is the reason why the training of CBF detectors usually

(a) Sample image displaying all labeled instances in an image.

(b) Cropped using aspect ratio standardization, yellow bounding box plus context.

(c) Cropped without using aspect ratio standardization, red bounding box plus context.

Figure 5.1: Sampling on Kitti pedestrian dataset.

involves several stages of bootstrapping (hard negative mining). The size of the subset, however, remains an empirical choice. Ideally, we seek two diverse and representative subsets for each class, positive and negative.

We revisit this data asymmetry problem and demonstrate an intuitive yet effective method to address it. We demonstrate that the CBF classifier is considerably more data hungry than commonly thought to be and that the quality and amount of the training samples have an important role in the training procedure. We propose a straightforward and efficient routine to collect suitable samples of the target class without the use of any data augmentation technique. We demonstrate that by providing suitable positive samples to the CBF classifier, the number of negative samples can be increased by a significant factor without incurring the risk of performance reduction or poor generalization. In this case, the CBF classifier can achieve an outstanding performance using solely HOG+LUV features without involving higher order features. Thus, our approach is orthogonal to methods improving through richer representation and consequently does not sacrifice any computational efficiency. We demonstrate that the proposed approach generalizes well through different datasets and different boosting methods, more concretely the discrete and real AdaBoost. Our implementation is based on the open source Computer Vision library [33].

The remainder of this chapter is organized as follows. Next, we review the most recent CBF-based methods in pedestrian detection and describe briefly our contribution. In section 5.2, we introduce our baseline detector and discuss in detail the settings that we change in the course of our experiments. We introduce the proposed approach and discuss how to exploit the training data in section 5.3. In section 5.4, we apply the proposed method on the test data, compare our results to state-of-the-art CBF-based methods, and demonstrate that the proposed data exploitation approach improves the detection accuracy significantly at virtually no additional computational expense. Finally, in section 5.5, we provide our conclusions.

5.1.1   Related work

CBF detectors rely fundamentally on the groundbreaking framework proposed by Viola and Jones [190]. A variant of the boosting algorithm [59, 162] is used to create a linear combination of decision trees and the cascade architecture rejects samples early from the classifier chain. Whereas the essential framework has remained unchanged, many aspects of it have evolved over time.

Unlike the popular deep learning approach, CBF-based methods cannot learn representations automatically from data and therefore require image features. Numerous feature types have been evaluated in the literature, such as SIFT [107], HOG [30], Fisher Vector [136], LBP [194, 192], color spaces [86], and local structure [191]. Among the mentioned features, the HOG descriptor, which is based on local image differences, is considered as the most popular and is frequently used in different approaches [30, 37, 50].

New features have been proposed to complement HOG [37, 8]. Possibly the most commonly used is ACF [34], consisting of ten channels. ACF is built on top of the basic integral channel features detector (ICF) [37] and simplifies the feature creation by aggregating neighboring pixel values into feature values instead of using Haar-like features. Numerous new features have been constructed by filtering the ACF with different filter banks, achieving remarkable performance gain [122, 210, 209, 208]. Current top performing CBF-based pedestrian detection methods, dominating Caltech [40] and Kitti datasets [60], are extensions of the ACF detector [34]. They achieve excellent performance by enriching the ACF via introducing increasingly more discriminative features. Recent works have even used CNN channels as features in combination with CBF classifiers and can, in this manner, combine the strengths of both approaches [202, 207, 19], which are highly discriminative CNN representation and fast cascaded tree-based proposal examination.

In addition to discriminative features, many other aspects of CBF detectors have been discussed in the literature, such as fast feature computation [34, 36, 153], acceleration of tree concatenation [3], and faster rejection of the negative samples through the cascade architecture [13, 35, 206].

The majority of the detection tasks in the field of computer vision are naturally cost sensitive, meaning that the number of targets to be detected are rare, e.g., pedestrian detection, where out of several thousand image patches only a small number contain a target. For these problems, the cost of missing a target is considerably higher than that of a false positive. One limitation of the commonly used boosting algorithms is that they minimize a quantity related to the classification error and thus, not necessarily decrease the number of false negatives. Adapting boosting to such detection domains has been a topic in the literature [175, 113]. [189] proposed an adaptation of the boosting algorithm called asymmetric AdaBoost such that the goal of each classifier in the cascade is not a low error, rather a high detection rate, and demonstrated an improved performance in the field of face detection.

5.1.2 Contribution

Within this study, we demonstrate that vanilla HOG+LUV features in CBF detectors have not yet saturated. We validate this by demonstrating that with additional control over the training samples, we can achieve similar or even improved performance compared with what well-engineered features provide, without decreasing the runtime of the detector for costly feature computation. In practice, we exploit the asymmetry of the dataset and demonstrate that the ordinary AdaBoost algorithm can not only manage a highly imbalanced training set but also that it actually benefits from a large and diverse amount of negative data. For this, we propose an effective pipeline that enables the CBF detector to collect a sufficient number of high-quality negative and positive samples without the use of data augmentation techniques and therefore omit the grid search for the correct parameters of such techniques.

We demonstrate that the proposed method, compared to the state-of-the-art CBF detectors, ranks as the second-best method on two challenging pedestrian benchmarks while running significantly faster. More importantly, the proposed method is orthogonal to the majority of the existing methods and can, therefore, benefit from integrating their concepts. In the remainder of this study, we refer to the proposed method as *insatiate CBF* (ICBF).

5.2 Ensemble

In this section, we describe, in detail, the settings for our experiment. We modify these throughout this study. Our baseline CBF detector is based on [34, 122].

5.2.1 Dataset

Dataset. We use the datasets introduced in section 3.2 for our experiments, namely the Caltech [39] and Kitti [60] pedestrian datasets.

Validation set. The training data in the Caltech dataset is divided into six sessions. The standard procedure in [40] suggests a 6-fold cross validation using five sessions in each phase for training and the 6th session for testing. We noticed that this setting was unfavorable for our experiments for two reasons: 1) Each of these divisions uses SI=30 and thus contains only a small number of images. The resulting training set is excessively small and cannot provide a sufficient number of diverse negative samples. 2) The log-average miss rate metric measures the detection quality in the range $[10^{-2}, 10^0]$ false positive per image. In this range, improvements in localization quality are not sufficiently representable. Because of 1) and 2), we perform our experiments on the Kitti evaluation set, which uses the average precision metric. For all our experiments, we use $\frac{3}{4}$ of the training data for the validation training and $\frac{1}{4}$ for the validation test. The divisions are performed in a manner where each of the four subsets contains approximately the same number of positive samples. In Kitti, the data does not derive from a stream, which means that there is a correlation in the divisions. We are not aware of any dividing procedure which avoids this. We report a posteriori our most interesting results on the standard test sets for comparison to the state of the art and to demonstrate that our findings generalize well.

5.2.2   Sampling

Collecting positive samples. To address the high diversity in the target classes, one commonly uses all positive samples that satisfy certain given criteria. For this, we use the height and occlusion ranges defined for the reasonable setting on Caltech and moderate difficulty on Kitti, see section 3.2. The ground truth of a sample locates a tight box that surrounds the object of interest entirely, see figure 5.1 (a) red bounding boxes. For training the CBF classifier, the underlying boosting algorithm 2.2 requires all samples to be represented by the same number of features, i.e., before feature creation, all cropped samples must be resized to the same predefined resolution. Figure 5.1 (b) and (c) display such positive patches, each containing a $\frac{64}{50} - 1$ margin in the vertical and $\frac{32}{20.5} - 1$ in the horizontal direction to include a context around the pedestrian. As can be observed, whereas the height does not vary considerably given a constant distance to the camera, the bounding box width can oscillate significantly with the pedestrian's pose (with the positions of the limbs, especially arm positions and relative angle). The main difference between the two patch series (b) and (c) is that (b) maintains the aspect ratio (ar) of the target object. For that, (b) uses an approach termed *aspect ratio standardization* that is suggested in [40] whereas (c) omits this step, as suggested in [125]. To apply the aspect ratio standardization either the height or width of the ground truth bounding box must be adjusted such that $\frac{b_w^g}{b_h^g} = ar$. ar can be regarded as a training hyperparameter that is set to be

$$ar \approx \frac{1}{|\mathcal{P}|} \sum_{i=1}^{|\mathcal{P}|} \frac{b_w^{g,i}}{b_h^{g,i}}. \tag{5.1}$$

The aspect ratio standardized bounding boxes can be observed in figure 5.1 (a) as yellow boxes. For sampling positives, we apply aspect ratio standardization and rescale the width of the ground truths bounding boxes while maintaining the height fixed. We further horizontally flip the cropped positive samples for a simple data augmentation to double their number.

Collecting negative samples. For negative sample mining, we use bootstrapping. Let $|\mathcal{N}|$ refer to the total number of negative samples used for the training and $\mathcal{N}'$ refer to the negative set used to create every intermediate monolithic CBF classifier during the bootstrapping stages. The initial negative set $\mathcal{N}_0$, with $|\mathcal{N}_0| = |\mathcal{N}|/2$, is collected by randomly sampling 25 negative samples per image from a subset of all available images. In each of the following bootstrapping stages, the same number of negative samples (i.e., $|\mathcal{N}_i| = |\mathcal{N}|/2$, where i denotes the ith bootstrapping stage) is collected and added to $\mathcal{N}' = \mathcal{N}_i \cup \mathcal{N}_p$, where $\mathcal{N}_p$ is a randomly chosen subset of $\mathcal{N}_0 \cup \mathcal{N}_1 \cup ... \cup \mathcal{N}_{i-1}$, with $|\mathcal{N}_p| = |\mathcal{N}|/2$. Negative samples must satisfy the criterion IoU < 0.1.

5.2.3   Model settings

Feature representation. We use the ten aggregated vanilla HOG+LUV features, as introduced in section 2.4.0.3, as the input to our CBF classifier without involving any other high order features.

Channel computation. Feature calculation is typically one of the most time-consuming operations in a multiscale detection pipeline. The input image is rescaled multiple times and the channels must be computed for each scale. We follow the scheme from section 2.4.0.3 and approximate seven out of eight scales per octave in the feature pyramid.

Template size. We investigate how the template and feature pool sizes affect the detection performance in terms of accuracy and computation time. Our initial tests on the validation set are performed using a template size of 64×32, where the pedestrian occupies a central area of 50×20.5, see section 5.2.2.

Model architecture. We use the constant soft cascade with constant rejection thresholds $\theta^* = -1$, see equation 2.4, and calculate the corresponding weights α_t of the decision trees using discrete AdaBoost, see algorithm 2.2.

Model capacity. The monolithic CBF architecture has a given maximum number of weak classifiers that are concatenated by the boosting algorithm. These weak learners (trees) have a defined maximum depth. The depth and number of the trees define the maximum capacity of the CBF classifier. We use 2048 weak classifiers of depth 2. We conduct bootstrapping three times to collect the final negative training set using $\{32, 128, 512\}$, where the numbers refer to the number of weak classifiers that constitute the CBF classifier in each bootstrapping stage, respectively.

Random forest. To accelerate the training process, the features are typically quantized into bins (256 in this work) and only a subset of the features is evaluated for every non-leaf node of the tree during the boosting training. Initially, we use $\frac{1}{16}$ of the total number of features, which is also used in [34]. It is worth mentioning that this randomization, similar to the dropout function in neural networks, facilitates regularizing the model and avoiding overfitting.

5.3 Exploiting the asymetry

5.3.1 Preliminary experiment

Our experiments are motivated by the observation that using the training procedure described in section 5.2, the performance of the ACF detector on the Caltech test set improves by less than 1 pp in MR when increasing the size of the training data by a factor of ten, which is performed by decreasing the SI value from 30 to three. The ACF+ detector [122] is a variant of ACF, our baseline detector, with the main differences that it uses more training data (SI=4 instead of SI=30 on Caltech) and deeper trees (depth five instead of two). In the case of the ACF+ detector, increasing the training data by approximately 33% (from SI=4 to SI=3) decreases the performance by approximately 1 pp in MR, see table 5.1. Note that, when enlarging the training set, we collect all positive samples available and retain the ratio between the number of negative and positive samples fixed, according to equation 5.2.

This discrepancy is not consistent with what is known in the literature. A large amount of high-quality data is fundamental to machine learning and

Method	SI		
	30	4	3
ACF	44.2%	-	43.7%
ACF+	-	29.8%	30.4%

Table 5.1: Performance in relation to the number of the training samples on Caltech test set.

supports training more robust models. [122] demonstrates that different model capacities require different numbers of training samples and that the performance can even decrease when using an excessive number of training samples. In this work, we argue that this could be related to the quality of the data and demonstrate in section 5.3.3 that indeed the quality of the positive samples has a relevant impact of the CBF classifier performance. Before beginning the experiments, we first define a hyperparameter to allow more control over the number of training samples for the experiments, the ratio (r), defined as

$$r := \frac{|\mathcal{N}|}{|\mathcal{P}|}. \tag{5.2}$$

For the task of pedestrian detection, it is common to use $|\mathcal{N}| \gtrsim |\mathcal{P}|$. Our baseline, the ACF detector, uses $r \approx 3$, whereas the ACF+ detector uses $r \approx 2$. We demonstrate through our experiments that the hyperparameter r is a subject deserving discussion. This is because of its direct effect on the weight distribution of the samples. The initial weight of each positive sample is r times the weight of each negative sample. Let $w_\Sigma = 1$ be the sum of the normalized weights of all samples during the boosting training. In the initial iteration of the training, $w_\Sigma/2$ is distributed among the samples of each class. This is a common practice that modifies the initialization from algorithm 2.2 to

$$w_i^1 = \frac{1}{2|\mathcal{P}|}, \quad \forall x^{(i)} \in \mathcal{P}, \tag{5.3}$$

$$w_i^1 = \frac{1}{2|\mathcal{N}|}, \quad \forall x^{(i)} \in \mathcal{N}, \tag{5.4}$$

see[188, 190]. Our aim is to feed the CBF classifier with as many negative samples as possible. By increasing the number of negative samples, we decrease their weights and simultaneously make the positive samples more cost sensitive, i.e., we use $|\mathcal{N}| \gg |\mathcal{P}|$ such that $\frac{w_\Sigma}{2 \cdot |\mathcal{N}|} \ll \frac{w_\Sigma}{2 \cdot |\mathcal{P}|}$.

The AdaBoost algorithm minimizes a quantity, ϵ, that is related to the classification error and thus does not necessarily decrease the number of false negatives. Therefore, each split is applied in a manner that the resulting ϵ is the lowest possible, where $\epsilon = \sum_i w_i \mathbb{1}[h(x^{(i)}) \neq y^{(i)}]$, with $w_i \in \mathbb{R}^+$ and $y^{(i)} \in \{-1, 1\}$ being the weight and label of the ith sample, respectively. In the case of random forest, only a subset of all possible splits is evaluated. The greater the weight of a sample, the greater its classification error becomes in the case it is misclassified. We expect that by using a high r value, we can influence the algorithm such that it first targets greedily negative samples. This is to avoid misclassification of the positive samples because false negatives are significantly more costly, higher ϵ, than false positives, initially r times costlier.

5.3.2   Greedily bootstrapping

As mentioned in section 5.2.1, we conduct our experiments on the Kitti
validation set. We train our detector using all the positive samples satisfying our predefined criteria while maintaining $r = 3$ fixed, see section 5.2.2.
This results in a training set consisting of 5328 positive and 15984 negative samples. The trained detector achieves an AP of 56.0%, which we
use as our baseline performance. The described bootstrapping strategy in
section 5.2.1 does not guarantee that negative duplicates are excluded, and
because a high number of negative samples is required for the next experiments, we modify the bootstrapping strategy in the following manner:
a new bootstrapping stage is added into the pipeline, changing it from 3
to 4 stages $\{32, 128, 512, 2048\}$. The output CBF classifier remains a linear
combination of 2048 weak classifiers. Before entering the last bootstrapping
stage, all negative samples collected in the earlier stages are discarded such
that the final CBF classifier trains using only the negative samples from the
last bootstrapping stage. Furthermore, to obtain a higher number of negative samples, we add a small positive shift $\Delta\alpha$ to each α_t in equation 2.4,
such that more samples can pass the threshold θ^*.

The collected negative samples are then ranked by their scores and the top
$36 \cdot |\mathcal{P}|$ ($r = 36$) samples with the highest scores, which we refer to as the
negative sample pool, are used for the following experiments. Moreover, we
change how the samples are collected. Rather than cropping them from
the RGB images and creating their channel features, we directly crop them
from the feature pyramid. In addition to being faster, this approach has
benefits that we explain in section 5.3.3.

At this point, the randomness in the training procedure affects two processes: how the negative training set, $\mathcal{N}$, is sampled from the negative sample pool and how the features are subsampled during the boosting training.
We evaluate different values of r to study its impact on the CBF classifier
performance. For every r value, we build three CBF classifiers varying the
seed of the random number generator $\in \{0, 1, 2\}$ and report the average AP
in table 5.2.

It can be observed that irrespective of r, using only hard negatives from the
last bootstrapping stage decreases the performance significantly. However,
increasing r appears to improve the performance. By increasing r from 3
to > 30, an improvement of approximately 10 pp is attained. However,
comparing this to the baseline performance, we have an overall reduction
of approximately 5 pp. Next, we repeat the above experiment, this time
following a new procedure to exchange the positive samples.

r	3	6	12	15	21	24	> 30
AP [%]	41.2	46.5	46.8	48.2	50.6	48.2	51.0

Table 5.2: Impact of the hyperparameter r on detection performance on Kitti validation test set. Horizontal flipping is used for data augmentation. Results are given in average precision (AP), higher is better. The negative samples derive from the newly added bootstrapping stage. Our baseline detector achieves an AP of 56.0% using $r = 3$.

5.3.3 Positive mining

Similar to the majority of current detectors, our underlying detector employs horizontal flipping to double the number of positive samples as a simple data augmentation approach. A recent study [125] demonstrates how effective a more sophisticated data augmentation technique can be. Existing pipelines can obtain a significant boost in performance when using an appropriate augmentation technique.

Typically, determining the best augmentation technique can be challenging owing to its dependence on multiple hyperparameters such as the template size, pad size, stride of the detection window, scales of the feature pyramid, and even the approximation of the channels. To avoid a grid search for determining the correct technique, we incorporate positive mining in the bootstrapping strategy, i.e., in the last bootstrapping stage we use the de-

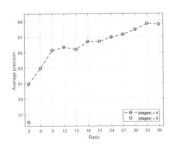
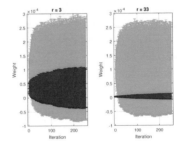

(a) |stage| = 4 refers to the use of samples from the newly added bootstrapping stage, where |stage| = 3 is our baseline CBF detector (AP = 56.0%) trained with $r = 3$ using the training protocol from section 5.2.

(b) Normalized weights of positive (green) and negative (red) samples at beginning of each AdaBoost iteration represented by their mean values and standard deviations. Displayed are only the first 256 iterations for $r = 3$ (left) and $r = 33$ (right).

Figure 5.2: Impact of the hyperparameter r on the CBF classifier.

tector to collect positive samples that have an IoU ≥ 0.5 and are unique (one proposal per ground truth). These samples are preferred by our CBF detector (with AP 56.0%) because they have survived NMS, meaning that they have received the highest score within all detections corresponding to the same instance (ground truth). The number of these samples depends on the true positive rate and can be less (or at maximum equal) to the number of the positive samples in the training set. Note that by adjusting θ^* ($\theta^* \to -\infty$, see equation 2.4), we can force the detector to have a true positive rate equal to one and thus make the number of samples equal to the number of the ground truths. However, we have observed a minimal impact on the performance when doing this and in favor of a faster runtime, we omit such a setup.

These samples have translations in position and scale and are affected by the channel approximation because they are directly collected from the feature pyramid, as mentioned in section 5.3.2. Moreover, as discussed in section 5.2.2, the aspect ratio standardization has an impact on the margin of the sample. Collecting positive samples as described has the effect of allowing the detector to choose the margin it prefers around the sample while the aspect ratio is maintained. It is worth mentioning that we also validated other criteria for collecting/creating positive samples. For example, we collected several instances before NMS and selected the one/ones with the highest IoU value/values. We also investigated an artificial method to produce jittered samples, however, using other approaches resulted, in the best case, in a slightly inferior performance.

Furthermore, [125] offers the interesting observation that flipping reduces performance because the enforced ratio standardization ($\frac{b_w^g}{b_h^g} = 0.41$) leads to non-aligned bounding boxes. We confirm this observation. Following this view, we remove the flipped samples and replace them with our collected samples with IoU ≥ 0.5. In figure 5.2, we repeat our experiments from table 5.2 using the newly created positive training set (GT $\cup$ IoU ≥ 0.5).

As can be observed in figure 5.2 (a), using the newly collected positive set, the performance improves by 5 pp for $r = 3$. Another 4 pp can be achieved when using $r = 9$. The next 4 pp are expensive to obtain; for this, we use $33 \cdot |\mathcal{P}|$ negative samples. The most remarkable observation is that, although the slope is no longer steep at higher r, the performance does not collapse under the quantity of negative samples. We confirm this observation in section 5.4. Figure 5.2 (b) illustrates the impact of the parameter r on the weights of the samples during the boosting training. As can be observed, the positive samples become relatively costlier for an increased value of r.

5.3.4   Model parameters

The AdaBoost algorithm operates in the sample $\times$ feature space. Each non-leaf node n outputs a binary decision by comparing a feature with a given index i against a threshold $\theta^t \in \mathbb{R}$ such that $h_n(x) = p_n \cdot sign(x[i_n] - \theta_n^t)$, where $p \in \{-1, 1\}$ is the polarity. By increasing the model capacity while maintaining the number of features constant, the boosting algorithm compares the same features repeatedly against different thresholds in different tree buildings. It is well known that more complex classification functions yield lower training errors, yet expose the risk of poor generalization [56]. It is also clear that maintaining both the number of features and the number of samples constant while increasing the model capacity can lead to overfitting. However, choosing $r > 30$ ensures that we have sufficient diversity in the training set to prevent the algorithm from learning specific features from the training set.

We increase our model capacity while holding the settings and training set the same as described in the previous experiments (see figure 5.2, $r = 33$). Table 5.3 summarizes our results. For the models displayed, the maximum number of non-leaf nodes in the entire CBF classifier is determined by $(2^{depth} - 1) \cdot |WC|$, where $|WC|$ refers to the maximum number of weak classifiers in the CBF classifier. Our deepest model compares in average every feature more than 99 times against different thresholds ($\frac{maximum-feature-requests}{feature-pool-size} = \frac{(2^5-1)\cdot 4096}{16 \cdot 8 \cdot 10} = 99.2$). Comparing this to state-of-the-art CBF classifiers that use several thousand features (see section 5.4.1 and [210, 19]) highlights the extent to which we can increase the density of our model without suffering poor generalization.

For the newly added last bootstrapping stage of our detector, we apply further changes as summarized in table 5.4. Although our baseline uses samples having their IoU with the ground truths less than 0.1 as negatives, we determine that setting the threshold for IoU to 0.5 decreases the localization error at lower FPPI ranges. This facilitates removing false positives

$|WC|$	depth $= 2$	depth $= 3$	depth $= 4$	depth $= 5$
2048	68.6%	72.0%	73.5%	74.0%
4096	69.9%	73.6%	75.3%	75.3%

Table 5.3: Average precision on Kitti validation test set while increasing model capacity. $|WC|$ refers to the maximum number of weak classifiers in the CBF classifier and depth denotes the maximum depth of the binary trees. Note that the first AP (68.6%) is given as a reference and was presented previously in figure 5.2.

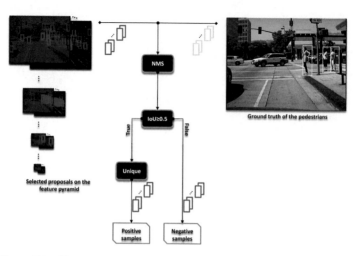

Figure 5.3: Illustration of the greedily bootstrapping routine. The routine can be divided into three phases. First, the CBF detector is applied to the RGB image. The proposals are collected using the corresponding locations on the feature pyramid. These proposals have a dimension of $\text{Ts}_h/\text{Ag} \times \text{Ts}_w/\text{Ag} \times |C|$, where Ts_h, Ts_w are the template height and width, respectively, Ag is the aggregation value, and $|C|$ is the number of the channels ($|C| = 10$ in the case of ACF). Second, NMS is applied to the corresponding RGB image coordinates of the proposals and suppresses, based on their score and overlap, redundant proposals. Surviving proposals with IoU ≥ 0.5 count as positive and the remaining proposals as negative samples. Finally, the positive proposals must be unique. This is because of the criteria of the benchmarks, see section 3.2.

with high scores around the target object. By applying the new IoU threshold to the last bootstrapping stage, we make use of all collected samples using one threshold to divide the samples into negatives and positives. As

Aspect	AP[%]
-	75.33
+ Negatives: IoU<0.5	78.07
+ Real AdaBoost	78.11
+ \Random forest	78.51

Table 5.4: Further fine-tuning of the parameter space. The IoU threshold is increased from 0.1 to 0.5. Instead of discrete AdaBoost, real AdaBoost composes the classifier chain without subsampling of the features.

in [122, 210], we observe that using more training data, real AdaBoost [59] rather than discrete AdaBoost improves the performance by a small amount. We observe the same behavior in our Caltech setup. Furthermore, we remove the randomness (\Random forest) of the training procedure without realizing any drawback and allow, in the last bootstrapping stage, for exhaustive searching over all features to determine the feature that offers the best split. It would appear that the number of diverse negative training samples regularizes the model and prevents overfitting.

Figure 5.3 summarizes and illustrates the greedily bootstrapping routine, which replaces the traditional data mining routine in the last stage of the bootstrapping. All detections are used for the training either as positive or negative samples. Collected samples, as discussed in section 5.3.2, are influenced by the channel approximation and the collected positive samples have translations in position and scale. The overall training time of our detector compared to our baseline is only extended by the training time of the last bootstrapping stage (all changes are performed only in the newly added stage, see section 5.3).

5.4 Experiments

In this section, we evaluate our CBF detector on the Caltech [39] and Kitti [60] test sets and present and discuss the obtained results. Furthermore, we investigate other important aspects of our detector that are loosely related to the above experiments.

Caltech. For the Caltech experiments, we set SI=3 and report our results in table 5.5. To distinguish the different versions, we compose their names in a fashion where their main differences are highlighted. In the following, Ag represents the aggregation value of the ACF planes, Ts the template size, and R refers to the used ratio according to equation 5.2. For further comparison, we address the different variants of our detector by their IDs, see tables 5.5 and 5.6, last column.

As mentioned in section 5.2.1, we repeat selected experiments posteriori on the test set. On the Caltech test set we attain the saturation point using $r = 21$, compare (1) and (2) in table 5.5. Because of the marginal improvement when increasing r from 15 to 21, for the following experiments we use a negative sample pool with $r = 15$ for faster training and less memory usage on our system (training Ag2,Ts64×32 or Ag4,Ts128×64 requires four times more memory than training Ag4,Ts64×32 for the same value of r).

Despite collecting samples directly from the feature pyramid, we notice that

Method	Aspect				ID
	\Approx	RoI	MR [%]	FPS [f/s]	
ICBF(Ag4,Ts64×32,R15)	-	-	25.2	21	(1)
ICBF(Ag4,Ts64×32,R21)	-	-	24.8	21	(2)
ICBF(Ag4,Ts64×32,R21)	✓	-	22.4	11	(3)
ICBF(Ag4,Ts64×32,R21)	✓	✓	20.1	14	(4)
ICBF(Ag2,Ts64×32,R15)	-	-	21.0	15	(5)
ICBF(Ag2,Ts64×32,R15)	✓	-	19.6	9	(6)
ICBF(Ag2,Ts64×32,R15)	✓	✓	17.2	11	(7)
ICBF(Ag4,Ts128×64,R15)	-	-	19.1	7	(8)
ICBF(Ag4,Ts128×64,R15)	✓	-	18.5	4	(9)
ICBF(Ag4,Ts128×64,R15)	✓	✓	16.7	5	(10)

Table 5.5: Results on Caltech test set using the reasonable setting. MR: log-average miss rate in the range $[10^{-2}, 10^{0}]$ false positive per image, lower is better. FPS: frames per second.

turning off the channel approximation improves our results, see \Approx in table 5.5. This is performed only for evaluation; the training procedure remains the same as discussed in section 5.3. Turning off the channel approximation is a costly step. To compensate for the runtime, we create a search space reduction as in [185, 28], see *region of interest* (RoI) in table 5.5. When using RoI, the computation performs exclusively in the region of the image where height $\in [90, 480]$ pixels, the image resolution on Caltech has height = 480 × width = 640 pixels. We furthermore examine the geometrical plausibility of the proposals using their heights and ground positions, see section 4.3.6.2.

The difference between (1) and (5) is their aggregation value. Decreasing this value results in a four times greater feature pool size, from $16 \times 8 \times 10$ to $32 \times 16 \times 10$. Clearly, using more features improves the performance. Increasing the feature pool size through the increase of the template size appears to perform best, compare (5) to (8). (5) and (8) have the same feature pool size. In the case of (8), the template size is increased to 128×64, where the pedestrian occupies a region of 100×41. In this case, because the minimum pedestrian height on Caltech is 50 pixels, the input image must be upsampled by one octave. Based on our setting, this adds eight new ACF planes to the feature pyramid, which is the reason why the runtime of the detector decreases significantly.

Kitti. The image resolution on Kitti varies by height $\in [370, 376]$ and width $\in [1224, 1242]$ pixels. For this reason, we do not apply our RoI module. The minimum height of the pedestrians in Kitti is 25 pixels, which means that

Method	Aspect				ID
	\Approx	RoI	AP [%]	FPS [f/s]	
ICBF(Ag4,Ts64×32,R36)	-	-	51.0	5.0	(11)
ICBF(Ag4,Ts64×32,R36)	✓	-	52.7	2.7	(12)
ICBF(Ag2,Ts64×32,R36)	-	-	54.1	2.9	(13)
ICBF(Ag2,Ts64×32,R36)	✓	-	54.8	1.8	(14)

Table 5.6: Results on Kitti test set using the moderate difficulty. AP: average precision, higher is better. FPS: frames per second.

for a template size of 64×32, the input image must be upsampled by one octave. For a template size of 128×64, we would have to upsample the input twice. The high image resolution and upsampling decrease the runtime of our methods on the Kitti dataset. Therefore, we train our model only with a template size of 64×32.

Because the number of pedestrians in the Kitti training set is considerably less than in Caltech, we can use a higher r value ($r = 36$). As can be observed, the results in tables 5.6 and 5.7 confirm our observation from figure 5.2, that the performance does not collapse under the quantity of negative samples. We achieve competitive results on Kitti, even without increasing the template size or applying our RoI module, see table 5.7.

5.4.1 Comparison to the state-of-the-art methods

Figure 5.4 compares our results to the state-of-the-art pedestrian detection methods. In both datasets, the first ranks are dominated by CNN-based methods. These methods commonly benefit from being pretrained on large datasets, a more sophisticated image representation, and expensive hardware to achieve adequate runtimes. As can be observed, the lightweight CBF-based methods continue to require improvement in terms of performance, to which our approach can contribute.

In table 5.7, we display a comparison to the currently leading CBF-based state-of-the-art methods on the Caltech and Kitti test sets. *Checkerboards* [210] and *CompACT* [19] (CompACT-Deep uses the VGG network to score the proposals where CompACT is the underlying CBF-based proposal generator) use similar strategies to improve the CBF classifier performance. Both methods enrich the proposals by introducing an extremely large set of new discriminative features.

Checkerboards uses a template size of 120×60 and an aggregation parame-

Method	Caltech		Kitti	
	MR[%]	FPS[f/s]	AP[%]	FPS[f/s]
ACF [34]	44.2	35*	39.8	5.5*
ACF+ [122]	29.8	15*	-	-
ICBF(Ag2,Ts64×32) (ID=7, 14)	**17.2**	**11***	**54.8****	**1.8***
ACF++ [125]	19.7	6	-	-
CompACT(GPU) [19]	18.9	4	54.9	0.75
LDCF++ [125]	14.9	2	-	-
Checkerboards [210]	18.5	0.5	54.0	-

Table 5.7: Comparison of performance, in terms of accuracy and runtime, with the state-of-the-art CBF-based methods on Caltech (MR metric) and Kitti (AP metric) test sets. Our result is labeled bold. Note that * refers to runtime measured on our system and ** to not including the RoI module.

ter of 6. The 10 ACF planes are each filtered with 61 predefined rectangular filters and thereafter input to a 4096 depth 4 trees. The reported performance is MR = 18.5% running at 0.5 FPS on Caltech and AP = 54% on Kitti, the runtime on Kitti is not reported.

CompACT extends the precomputed ACF planes by different feature types (such as self-similarity [167], Checkerboard [210], locally decorrelated HOG [122], CNN, and CNN Checkerboards (CNNCB)) resulting in an extremely large feature pool size. To accelerate the computation, the majority of the features are only created on demand and GPU-computation is involved to generate the CNN features. Moreover, a complexity-aware variant of the AdaBoost algorithm manages the arrangement of the weak classifiers to achieve an acceptable trade-off between runtime and accuracy. The reported runtime is 4 FPS on Caltech while achieving MR = 18.9% and AP = 54.9% on Kitti at 0.75 FPS.

Ohn-Bar et al. [125] demonstrate the importance of data augmentation. Using the same setting as our ICBF(Ag4,Ts128×64) (ID=9), they report a performance of MR = 20.47% using 200000 negative samples and a scale and crop technique to create a positive training set with a size equal to four times the number of available ground truths. When using context analysis (see table 5.7, ACF++), a performance of 19.71% is reported. On our system, we obtain for ACF++ the same runtime as for ID=9 (4 FPS); however, we report the runtime from their work. Furthermore, LDCF++ applies four LDCF filters [122] on top of ACF++ resulting in a feature pool size enlarged by a factor of four. Hence, the runtime decreases by approximately the same factor with a performance gain of ≈ 5 pp (also, in the case of LDCF++, we obtain a different runtime of ≈ 1 FPS on our system). We emphasize that

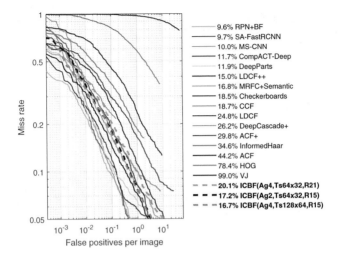

(a) Caltech test set, reasonable setting.

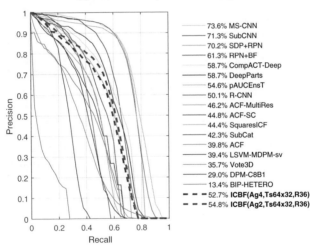

(b) Kitti test set, moderate difficulty.

Figure 5.4: Comparison to state-of-the-art pedestrian detection methods. Our results are shown in dashed lines labeled bold.

we do not use any data augmentation technique for the positive training samples; rather we focus mainly on how to exploit the asymmetry in the dataset and improve through increasing the number of negative samples.

Comparing our results to CBF-based state-of-the-art detection methods certainly proves the effectiveness of the suggested changes. We are not aware of any CBF-based detection method that can achieve a superior performance using such a small feature pool, i.e., $16 \times 8 \times 10$ features per proposal. Most importantly, the proposed approach is orthogonal to the mentioned methods and can, therefore, benefit from integrating those concepts. Our fastest variant (ID=2) can also compete with methods specially designed for a well-balanced accuracy and speed trade-off. One of the fastest methods on Caltech with a high accuracy, which has recently been reported, is *DeepCascadeED* [2]. DeepCascadeED achieves MR = 26.2% running at 15 FPS on a GPU. It attains a detection rate of < 80%, see figure 5.4. This means that the majority of the proposals do not pass the entire cascade to be rescored by the computationally most expensive network arranged at the end of the cascade. ICBF(Ag2,Ts64×32,R15) (ID=5) achieves the same runtime on a moderate CPU providing a detection rate of $\approx 95\%$ and > 5 pp better performance.

5.5 Epilogue

We revisited the asymmetry in a dataset and investigated its impact on the AdaBoost algorithm. Based on our investigation, we presented a novel approach that focuses on exploiting the asymmetry by introducing simple and efficient changes within the bootstrapping procedure. These changes enable the detector to collect a sufficient number of negative and high-quality positive samples. The resulting training set is highly imbalanced and managed extremely well by the classifier. Consequently, our detector, which is based on the cascaded boosted forest classifier, has demonstrated the ability to increase its discriminative power, resulting in one of the fastest and best-performing descendants of the channel feature detectors family. In particular, one of the trained models achieved the second-best rank on two challenging datasets among related detectors while running significantly faster. Moreover, the presented approach can be easily integrated into other boosted-based training procedures where negative data mining is crucial.

Chapter 6

Conditional Multichannel Generative Adversarial Networks
A Novel Approach for Generating Synthetic Traffic Signs

6.1 Motivation

Generative adversarial networks (GANs), proposed by Goodfellow et al. in [71], are considerably new in the field of artificial intelligence and yet a topic of many studies in the recent past. One of the main reasons that make GANs popular is that they have proven to produce visually superior representations than other generative methods such as variational autoencoders [89] and autoregressive models [126]. In addition to their representational ability, GANs have demonstrated a potential for solving real problems in different branches of research, see [131, 143, 102, 87].

The adversarial modeling frameworks are typically built using two main models: the *generator* that creates the representation and the *discriminator* that guides the generator to do so. To the best of our knowledge, all frameworks use a fundamentally similar construction for both models. That is, the discriminator and the generator use the same kernels to process multiple channels of the input image. This functions well in cases where both channels have a similar texture and the same resolution, for example, when the input channels originate from a regular RGB camera chip.

Within this work, we demonstrate that in cases when the above precondition is not fulfilled, the ordinary architecture within such frameworks struggles to create appropriate representations. An example of where this precondition is not fulfilled is when the underlying sensor provides a red-clear-clear-clear raw image. We propose a general solution for such cases that does not require exhaustive tuning of the model's hyperparameters and demonstrate the enhanced representation ability of the proposed framework on traffic sign representations with samples captured from a camera with the mentioned pixel topology. Our implementation is based on the open source CNN library [187].

The remainder of this chapter is organized as follows. In the balance of this section, we review relevant studies focusing on the use of synthetic traffic signs, the most recent works in the domain of GANs, and describe briefly our contribution. In section 6.2, we describe the input signal, its composition, and the initial architectures of the proposed adversarial models. We also demonstrate that the proposed architectural changes facilitate the improvement in the quality of the generated samples significantly without requiring exhaustive hyperparameter tuning. In section 6.3, we demonstrate the relevant characteristics of the synthetic data and the proposed pipeline in more depth. Throughout this section, we discuss interesting usage scenarios for the generated traffic samples. Finally, in section 6.4, we provide our conclusions.

6.1.1   Related work

In this section, we discuss the use of synthetic data in the field of traffic sign recognition and review the most recent and relevant works on GANs.

6.1.1.1   Traffic sign recognition and synthetic data

Traffic sign recognition has a high relevance in autonomous driving. Owing to the rigid and planar appearance of the object of interest, the task can be considered as less challenging compared to other recognition disciplines. However, the difficulty in determining a global solution in this field comes from other aspects. For example, the German Traffic Sign Recognition Benchmark [173] is considered one of the most popular and challenging benchmarks available online. The dataset contains approximately 40 different traffic signs cropped from more than 50000 images. In 2012, Ciresan et al. in [26] proposed the use of a combination of 25 CNNs that can outperform human accuracy by a factor of two. However, the proposed solution applies only to a portion of the existing traffic signs on German streets. Considering that each country uses a different type of traffic sign exemplifies the requirement for more advanced methods and synthetic data to reduce the effort behind the data preparation.

In [119], the authors apply a series of geometric and photometric operations (rotations, hue and lighting variations, and background transplants) on traffic sign templates to create synthetic samples. These samples are used to train a decision tree-based detector. The authors, however, report inferior results compared to when using real data.

Conversely, [24] jointly uses real and synthetic data created from pictograms. The authors report improved results of the large-scale detection and classification pipeline compared to using solely real samples. To create the synthetic data, they use a similar solution to [119] and vary H and V values in the HSV color space, apply rotation, scaling, shifting, salt and pepper noise, and other techniques.

In [120], the authors advance this concept and achieve comparable results for classifiers trained on real and synthetic data using pictogram-based synthetic samples.

Recently, Haselhoff et al. in [73] have proposed a more advanced solution to generate synthetic traffic signs using a Markov random field and two underlying real samples. The solution exchanges the background and foreground of two real samples and increases, therefore, the diversity within the training set. In experiments, portions of synthetic data substitute the same amount of real data in the training set. The results demonstrate that using a large amount of synthetic data can improve the classifier performance significantly when real samples are scarce.

6.1.1.2 Generative adversarial networks

GANs have been an exciting research field and topic in many different works in recent years. In [140], the authors demonstrate that training a discriminator enables the extraction of semantically meaningful features. The trained discriminator can then be used on a different dataset to fulfill a classification task without any significant adaptation and further supervised training. Furthermore, [41] and [140] indicate that specific filters of a trained generator have learned to draw specific objects such that it is possible to trigger them on demand by changing (interpolating) the noise vector accordingly.

In [118], the authors demonstrate that by feeding the generator and discriminator with conditions, they can create MNIST digits conditioned on class labels. In [124], the idea is further evolved by introducing an additional auxiliary classifier that contributes to the training procedure.

In [83], the authors use an image instead of a noise distribution as an input
to the generator and present a pipeline that can translate an image into a
different one.

GANs have also been successfully used for classification tasks. Zheng et
al. in [212] use generated images of people for the task of person reidenti-
fication. To accomplish this, the authors require all output neurons to fire
weakly when the input to the classification network is a generated sample.
The reason for this is that the generated samples do not actually appear
human-like and therefore, can be considered as hard negatives. Wang et
al. in [193] apply an opposite approach and use GANs to generate samples with occlusions and deformations that are rare in the dataset and in
this manner, provide hard positives for the classifier training. Sixt et al.
in [169] use GANs to make images from a 3D model appear more realistic
and demonstrate that the use of the additional synthetic data improves the
classifier performance significantly.

Probably the most successful employment of GANs has been demonstrated
in the domain of semi-supervised learning. In [123, 170, 96, 158, 29], impressive results are realized using a small subset of the labeled real data
for supervised training and the remaining portion of the data for the unsupervised training of the adversarial networks. Interestingly, [158] and [29]
state that for such pipelines, the generated data are not required to appear
visually "good"; they actually encourage using "bad" looking samples for
improving the classification performance.

6.1.2 Contribution

Within this study, we propose a general adaptation of GANs into domains
where the input image has multiple channels with inherently different textures and resolutions. To achieve this, we use several known concepts in the
field of deep learning. More concretely, we use weight sharing for all layers
that are responsible for high-level semantics in both models, the discriminator and generator. This allows the generator to create the representations
of separated, multiple channels using the same seed noise vector. Further,
this enables the representations to jointly learn through the same loss function (or multiple loss functions). At layers that are responsible for low-level
details, we separate the kernels. This allows individual kernels to become
specialized in generating (for the generator) and extracting (for the discriminator) characteristic features from the different channels. Moreover,
we propose architectural changes for both models that improve the quality of the representations, including a more consistent composition for the
generator network.

We confirm that the proposed solution does not require exhaustive tuning of the model's hyperparameters and demonstrate the enhanced representational power using traffic sign samples that are captured by a camera with a red-clear-clear-clear pixel topology. Our solution is sufficiently general such that any adaptation of the concept of conditions (continuous and discrete) are easily accepted into the framework. This is critical for traffic sign representations and has practical features that we discuss throughout this work.

As an additional contribution, throughout this study, we discuss research that has successfully employed synthetic data in the field of traffic sign recognition. We analyze our generated synthetic traffic signs in detail and suggest, based on their features and related work, alternative solutions on how to employ our framework for classification purposes.

6.2 Evolving the modeling framework

6.2.1 Input signal

In autonomous driving, cameras are frequently used that do not have a regular RGB pixel topology; rather they use a red-clear-clear-clear pixel arrangement. This kind of filter provides a higher light intensity than the RGB filter. In this case, red-clear-clear-clear indicates a pixel with a red color filter and three neutral color (clear) filters. The arrangement of the pixels provided by such a sensor is displayed in figure 6.1 (left). Such input signals must first be pre-processed. In our experiments, we determine that this pixel topology can be better processed, i.e., result in more robust classifier models, when we separate the channels, see figure 6.1 (right). This separation creates two channels, gray and red, having resolutions of $H \times W$ and $\frac{H}{2} \times \frac{W}{2}$, respectively, where H and W are the height and width of the input signal, respectively. It should be mentioned that after the separation, missing gray pixel values are interpolated and then the gray channel is smoothed. The number of the gray pixels before interpolation is three times the number of red pixels; after the interpolation, it becomes four times this number.

To use the traffic sign samples in an adversarial framework, we must first define the input resolution of the samples. We choose H and W to be 32 pixels for both channels, resize them using bilinear interpolation, and arrange them into a matrix with the dimension of $32 \times 32 \times 2$, where 2 denotes the number of channels. This arrangement is required by any proposed framework of which we are aware. For the following experiments, we use this input size unless explicitly stated otherwise. Furthermore, we project each sample linearly into the range of $[-1, 1]$ to match the output nonlinearity in our generator, which is a tanh function as displayed in table 6.1. For this, we use the global minimum and maximum values of the data format used (unsigned integer with eight bits). The alternative to this would be using the minimum and maximum values of each sample individually, which then produces two free to choose parameters when we project the generated samples back into the $[0, 255]$ space.

6.2.2 Generative adversarial networks

GANs consist of two main components, a discriminator (D) and a generator (G). Both models, D and G, can be any differentiable function and are typically implemented either as neural networks or convolutional neural networks. The objective of G is to synthesize samples that mimic real samples whereas that of D is to discriminate real samples from generated samples. Both models are trained jointly and therefore use only the gradients of the error function that derive from D. This is because G does not have its own objective function.

Let z denote a d-dimensional random vector ($z \in \mathbb{R}^d$) drawn from a uniform distribution with components within the range of $[-1, 1]$. G accepts z as an input and outputs $x' = G(z)$, with x' having the same dimension as the sample from the real data, x. We denote the distribution of the synthetic samples, $G(z)$, as p_G and those of the real data as p_x. We train D to respond to samples from the real data distribution with $D(x) = 1$ and to the generated samples with $D(x') = D(G(z)) = 0$. G, conversely, uses the gradients of D's error function to learn how to "fool" D such that the output $D(G(z))$ approaches 1. This training routine corresponds to a min-

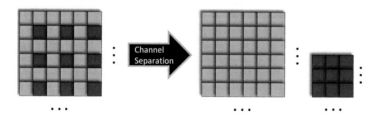

Figure 6.1: Pixel topology of a red-clear-clear-clear camera sensor (left) and the preparation of the input signal, i.e., channel separation and smoothing of the gray channel (right).

max two-player game and is played via solving

$$\min_{G} \max_{D} V(D, G), \tag{6.1}$$

where

$$V(D, G) \equiv \mathbb{E}_{x \sim p_x} \left[\log \left(D(x) \right) \right] + \mathbb{E}_{z \sim p_z} \left[\log \left(1 - D(G(z)) \right) \right], \tag{6.2}$$

and p_z denotes the distribution that is used to draw z. Both models pursue contrary goals and the solution to this game is called the *Nash equilibrium*. In practice, the problem formalized above is solved by learning the parameters of D and G, w_D and w_G, respectively through the gradient update

$$w_D^{t+1} = w_D^t + \mu \nabla_{w_D} V(D^t, G^t), \tag{6.3}$$

$$w_G^{t+1} = w_G^t - \mu \nabla_{w_G} V(D^t, G^t), \tag{6.4}$$

where μ is the global learning rate, t is the iteration number, and $\nabla_w \equiv \frac{\partial}{\partial w}$, see equation 2.8. The opposite goals of the two models are converted to opposite directions (see $+$ and $-$ in the gradient update rule) to where the gradients of D's error function point. That is, D is trained to maximize $D(x)$ and minimize $D(G(z))$ and G is trained to maximize $D(G(z))$ and while doing so, concurrently decrease the confidence of $D(x)$.

Goodfellow et al. in [71] demonstrate that, given a sufficient number of training iterations and capacity of both models, the distribution p_G converges to p_x. This means that the generator learns to synthesize images from the random noise vector z that resemble, from the perspective of D, images drawn from the real data distribution p_x. Consequently, D becomes unable to distinguish between the two distributions and begins guessing at random with $D(x) = D(G(z)) = \frac{1}{2}$ unable to determine any differences.

6.2.3 Architectural design

Training adversarial networks can be challenging. Unlike many optimizing problems in the vision domain, we do not seek a (global) minimum of a cost function; rather, we seek the Nash equilibrium of the loss function given in equation 6.1. The progress of one model can undo the progress of the other. To date, to the best of our knowledge, there is no fundamental answer to the question of how to train such a pipeline in the best manner. However, solutions are proposed in [32, 140, 111, 5, 70, 4] to stabilize the training procedure. Radford et al. in [140] propose a framework that is based on deep CNNs and provides architectural guidelines for both networks that have been demonstrated to function well.

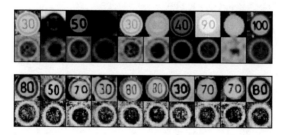

Figure 6.2: Top: real training samples. Bottom: generated traffic signs using the proposed initial models. In both, the first row corresponds to gray channels and the second row to red channels.

For our initial experiment, we create a CNN-based GANs pipeline that we maintain architecturally close to the guidelines from [140]. We use an equal number of convolution and transposed convolution layers in D and G, respectively. We furthermore use a batch normalization layer [82] after each convolution and transposed convolution layer to accelerate the training process, leaky ReLU with a slope of 0.2 as nonlinearity in D and ReLU in G, except for the last layers. Layers that create the score (in D) and the output dimension (in G) use sigmoid and tanh nonlinearities that project their outcome into the range of $(0, 1)$ and $(-1, 1)$, respectively. For the training, we use the Adam optimizer [88] with momentum terms $\beta_1 = 0.5$, $\beta_2 = 0.999$, and $\mu = 0.00002$. Moreover, we include changes for adaptation to the new domain. To create an output resolution from the noise vector z, Radford et al. in [140] suggest a linear projection of z ($z \in \mathbb{R}^{100}$) and a reshaping of the outcome of this operation afterward. We use instead, a noise vector $z \in \mathbb{R}^{1 \times 1 \times 100}$ and directly apply the transposed convolution on it to create the output resolution. Therefore, the reshaping operation becomes obsolete. Both approaches are linear and therefore mathematically equivalent, yet using transposed convolution makes the composition of the generator network more consistent. The generator can thereafter be built by concatenating several blocks, each consisting of one transposed convolution, one batch normalization, and one nonlinearity layer.

We notice that when using filters with a kernel size of 5×5, the generator creates blurry samples. Reducing the receptive field of the filters to 3×3 for kernels in both the convolution and transposed convolution layers improves the quality of the samples. The zero-padding parameters are accordingly adjusted such that the outcome resolution of each layer remains the same. This is performed in all layers except for the first of G and the last of D, which have 4×4 filters. We observe that this approach provides a superior

Layer	Convolution	BN	Nonlinearity	Outcome
1	64@ $3 \times 3 \times 2$ Stride 2	✓	Leaky ReLU Slope 0.2	$16 \times 16 \times 64$
2	128@ $3 \times 3 \times 64$ Stride 2	✓	Leaky ReLU Slope 0.2	$8 \times 8 \times 128$
3	256@ $3 \times 3 \times 128$ Stride 2	✓	Leaky ReLU Slope 0.2	$4 \times 4 \times 256$
4	1@ $4 \times 4 \times 256$ Stride 1	-	Sigmoid	$1 \times 1 \times 1$

Layer	Convolution^T	BN	Nonlinearity	Outcome
1	512@ $4 \times 4 \times 100$ Stride 1	✓	ReLU	$4 \times 4 \times 512$
2	256@ $3 \times 3 \times 512$ Stride 2	✓	ReLU	$8 \times 8 \times 256$
3	128@ $3 \times 3 \times 256$ Stride 2	✓	ReLU	$16 \times 16 \times 128$
4	2@ $3 \times 3 \times 128$ Stride 1	-	Tanh	$32 \times 32 \times 2$

Table 6.1: Detailed architecture of both models (D top, G bottom). Convolution^T denotes the transposed convolution operation. Convolutions and transposed convolutions are described by N@$H \times W \times C$, where N, H, W, C are the number of neurons, height, width, and depth of the kernels, respectively. BN represents batch normalization. Outcome displays the output resolution after each operation block (Layer).

representation for small input sizes. We also set the hyperparameter $T_D = 1$, as suggested in [71]. Further, we introduce a second hyperparameter named T_G, indicated in algorithm 6.6. This is because we observe that D becomes overly confident regarding the input early in the game. To improve the progress of G, we use $T_G = 2$ and thus update G twice during one training iteration.

Using this setting, the proposed models can generate reasonably acceptable representations of the gray channels, yet repeatedly fail to address the high frequency of the red channels, as can be observed in figure 6.2. Our assumption that this could be related to the use of unoptimized hyperparameters has proven to be incorrect. In the following section, we present a different approach where we retain the above settings and change the architectures of our models.

6.2.4   Multichannel GANs

In the information flow in a CNN, the first layer typically decodes low-level details and higher layers decode high-level semantics. This is the case in our discriminator as it is an ordinary binary classifier. In the case of our generator, the information flows in the opposite direction. That is, low layers decode abstract semantics from the random noise vector, whereas higher layers manifest the details. This means that the interaction between both models is highly influenced by layers that are accountable for low-level details.

The first layer of our D model, as presented in table 6.1, has 64 neurons. Each of these neurons has a kernel size of $3 \times 3 \times 2$ and therefore, its activation depends on a 3×3 united region of both input channels, gray and red. Hence, the question is: what happens if the texture of these unified regions is highly dissimilar?

All frameworks that we are aware of can (or can be adapted to) process multiple input channels in cases where all the channels have the same dimensions. As indicated in table 6.1 and figure 6.2, our baseline models use a setup as described above and struggle to generate a high-frequency red channel and a low-frequency gray channel simultaneously. To solve this, we use the concept of weight sharing. In both models (D and G), we share weights in all layers that are accountable for abstract high-level semantics between both channels. For the weights at layers that create/extract the low-level details, we choose individual weight values for the different channels. In this manner, the different kernels of both networks can become specialized in extracting and generating specific attributes from each channel. Furthermore, the separation allows us to use two channels of different sizes as inputs. We use the red channel with a resolution of 16×16 and the gray channel with 32×32. This is because that after separating the channels, the number of gray pixels is four times the number of the red pixels, see figure 6.1.

Figure 6.3: Generated traffic signs using the proposed MCGANs. The first row corresponds to gray channels and the second row to red channels.

To implement these changes, we remove the first layer of D and replace it with two input layers, each with 32 neurons to make D capable of processing two separated input channels. The stride of the layer that processes the gray channel is set to two; that for the red channel to one. The outcomes of both convolutions have the same resolution ($16 \times 16 \times 32$), are separately batch normalized, input to a leaky ReLU nonlinearity, and afterward united to continue jointly the forward pass. In this manner, both branches (gray and red paths within the architecture) have access to the same loss function. For the error backpropagation, we divide the gradients of D's error function at the location where the outcomes of both branches are united and train each branch separately.

For G, we perform a similar operation, however, conversely. The fourth layer is replaced with two new layers each with a $3 \times 3 \times 128$ kernel. These account for gray and red channels and have strides equal to two and one, respectively. Both layers receive the outcome from the third layer as an input and can thus be triggered by the same noise vector and any condition to which it is linked. The gradients of D's error function that derive from the separated branches are summed before the third layer and back-passed to the shared layers of G.

The proposed architecture, that we refer to as *multichannel generative adversarial networks* (MCGANs), is illustrated in figure 6.4 and the outcome of the new generator is displayed in figure 6.3. As can be observed, this separation of the kernels solves the problem of the joint receptive field at layers accountable for low-level details. The generator is now able to create an appropriate representation of the red channel. Further, this separation noticeably improves the representation quality of the gray channel.

6.2.5 Conditional MCGANs

The use of conditions in GANs is becoming more popular owing to several benefits that they provide. GANs are capable of generating samples of different classes; however, they do not offer any solution on how to trigger G to create a sample of a predefined class. Furthermore, the use of additional information such as labels makes the training more stable and even faster because we additionally use gradients that derive from a different objective function.

Our networks, as displayed in figure 6.4, share higher layers and the objective function in D for both representations (gray and red). This allows extending the framework by additional objective functions. Inspired by [124], we create an auxiliary classifier (AC) on the top of our D architec-

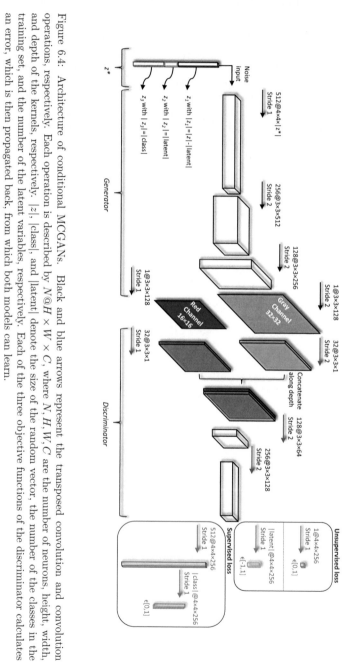

Figure 6.4: Architecture of conditional MCGANs. Black and blue arrows represent the transposed convolution and convolution operations, respectively. Each operation is described by $N@H \times W \times C$, where N, H, W, C are the number of neurons, height, width, and depth of the kernels, respectively. $|z|$, $|class|$, and $|latent|$ denote the size of the random vector, the number of the classes in the training set, and the number of the latent variables, respectively. Each of the three objective functions of the discriminator calculates an error, which is then propagated back, from which both models can learn.

Algorithm 6.6 *Mini-batch training of conditional MCGANs*

Input:
- T number of training iterations
- G generative model
- D discriminative model
- $\{(x^{(i)}, y^{(i)})\}_{i \leq N}$ real samples and their corresponding labels
- p_z noise distribution
- B_s mini-batch size
- T_G, T_D hyperparameters

Output:
- G, D

1: **for** $t = 1, \ldots, T$ **do:**
2:     **for** $t_D = 1, \ldots, T_D$ **do:**
3:         Draw B_s noise vectors, $\{z^{(1)}, \ldots, z^{(B_s)}\}$, from p_z
4:         Draw B_s pairs, $\{(x^{(1)}, y^{(1)}), \ldots, (x^{(B_s)}, y^{(B_s)})\}$
5:         Update the discriminator by ascending its stochastic gradient

$$\frac{1}{B_s}\nabla_{w_D} \sum_{i=1}^{B_s} \left[\log\left(D(x^{(i)})\right) + \log\left(1 - D(G(z^{(i)}))\right)\right]$$

6:         Update the discriminator by descending its stochastic gradient

$$-\frac{1}{B_s}\nabla_{w_D} \sum_{i=1}^{B_s} \sum_{j=1}^{|class|} y_j^{(i)} \log\left(z_{3_j}'^{(i)}\right)$$

$$\frac{1}{2B_s}\nabla_{w_D} \sum_{i=1}^{B_s} \sum_{j=1}^{|latent|} \left(z_{2_j}^{(i)} - z_{2_j}'^{(i)}\right)^2$$

7:     **end for**
8:     **for** $t_G = 1, \ldots, T_G$ **do:**
9:         Draw B_s noise vectors, $\{z^1, \ldots, z^{B_s}\}$, from p_z
10:       Update the generator by descending its stochastic gradient

$$\frac{1}{B_s}\nabla_{w_G} \sum_{i=1}^{B_s} \log(1 - D(G(z^{(i)})))$$

$$-\frac{1}{B_s}\nabla_{w_G} \sum_{i=1}^{B_s} \sum_{j=1}^{|class|} z_{3_j}^{(i)} \log\left(z_{3_j}'^{(i)}\right)$$

$$\frac{1}{2B_s}\nabla_{w_G} \sum_{i=1}^{B_s} \sum_{j=1}^{|latent|} \left(z_{2_j}^{(i)} - z_{2_j}'^{(i)}\right)^2$$

11:     **end for**
12: **end for**

---

ture that shares all layers with the binary classifier of D, except for the fully connected layer (FC). The AC has two new FC layers with 512 and $|class|$ units, receptively, where $|class|$ is the number of classes in the training data. This extension is beneficial, especially in the case of traffic sign representations, where we wish to know the class membership of the generated samples. We trigger our G model to create samples of a specific class by linking the label of the desired class as additional information to the input noise vector. Thus, $x' = G(z)$ evolves to $x'(y) = G(z, y)$, where $y \in \mathbb{N}^{|class|}$ is the class label given as a one-hot vector. The AC is trained to minimize a loss function, which we choose to be the cross entropy with the form $\mathbb{E}[-\sum_j y_j \cdot \log(z'_{3_j})]$, with z'_3 being the outcome of the last FC layer. The nonlinearity of this layer is a softmax function, thus $\sum_{j=1}^{|class|} z'_{3_j} = 1$. During training, the added classifier learns in a supervised manner from real samples, i.e., real samples are used as input to D and the real labels are used as y. For the generated samples, we create random one-hot labels z_3 that we also use as input to the generator.

We further extend the proposed pipeline by a third loss function, mean squared error, given as $\mathbb{E}[(\sum_j z_{2_j} - z'_{2_j})^2]$, where z_2 is a vector sharing its components with z and z'_2 denotes the output of a new FC layer. We create the new FC layer with the same input dimension as that for the binary classifier, however, with a different number of output neurons ($|z_2|@4 \times 4 \times 256$) and remove the sigmoid nonlinearity that would have come after the FC layer, otherwise $z'_2 \in [0, 1]$ while $z_2 \in [-1, 1]$. In our implementation, we choose z_2 to have 3 components, $|z_2| = 3$. This approach was proposed in [22] and allows the learning of latent variables from the z vector in an unsupervised manner. Further, as discussed in [22], the new loss function makes the training more stable, helps the pipeline to converge faster, and has virtually no additional computational cost. Henceforth, we name the new composition of the noise vector as z^*, see figure 6.4. Algorithm 6.6 summarizes the MCGANs training routine.

6.3   Experiments

In this section, we examine the quality of the synthetic samples and discuss their features.

6.3.1   A closer examination of synthetic traffic signs

The use of traffic signs offers unique opportunities to explore how the proposed MCGANs create the synthetic data. All real samples share similarities. More concretely, they all include a random background, a circular foreground, a red circle surrounding the actual content of the sign, and only differ in the small area that includes the actual sign content. We know that when we remove z_2, z_3, and the corresponding loss functions, G is capable of learning to generate all samples from the training set. Next, we explore the contribution of the conditions more closely.

In figure 6.5 (a), we fix z in each row and change z_3 along the columns. As can be observed, the surroundings of the actual content, perspective, and light conditions are virtually unchanged, whereas varying z_3 changes the actual content of the traffic sign. It would appear that G initially generates a template and uses the given label z_3 to fine-tune the template and capture the p_x distribution.

In figure 6.5 (b), we indicate how samples, which use the same z, degenerate to the same template when we downscale their one-hot label z_3 along each row with the factor $f_s \in [0, 1]$. This confirms our observation above. Such use of discrete conditions could be interesting to generate hard negatives and complement the training data. For this, discrete conditions can be employed to create these template-like samples with natural-looking surroundings. The goal would be to encourage the classifier to focus only on information necessary to predict the class label and become invariant to unimportant details.

In figure 6.5 (c), we demonstrate transferring the appearance of generated samples along the row. Therefore, we maintain z fixed and provide G with a new label that is not a one-hot vector and given as $z_3''^{(i,i')} = z_3^{(i)} \cdot f_s + z_3^{(i')} \cdot (1 - f_s)$, where $z_3^{(i)}$ and $z_3^{(i')}$ are one-hot labels of two different classes. As can be observed, G can smoothly transfer one sign into another solely through the application of the new label.

In figure 6.5 (d)-(f), we indicate the impact of the learned latent variables of z_2. The vector z_2 has 3 components that have learned the most incisive attributes of the training samples. Through the components of z_2, one can respectively influence the background illumination, the scale, and the overall illumination of the generated samples as displayed in (d), (e), and (f). The use of such a technique could be interesting in the field of traffic sign recognition where trained models must typically address large variations in visual appearances due to illumination changes, partial occlusions, rotations, and weather conditions. For this, a trained generator must be short-circuited to a trained classifier. For each generated sample, analyzing the score determined by the classifier and the corresponding continuous condition allows for identifying the model's weaknesses.

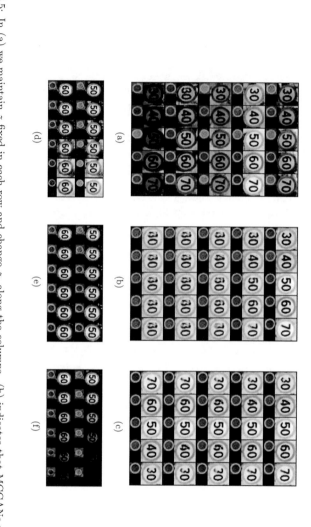

Figure 6.5: In (a) we maintain z fixed in each row and change z_3 along the columns. (b) indicates that MCGANs use z to create a template for all samples. (c) displays (from top to bottom) how each sample can be transferred solely by changing z_3, and (d)-(f) display the impact of the three components of the learned latent variable z_2. Gray channels are displayed in a resolution of 32×32 and the corresponding red channels in 16×16.

6.3.2 Quality of the generated samples

One of the problems that continues to be inadequately solved when discussing representation learning is the missing metric for determining if the quality of the generated data is sufficient. Moreover, the lack of such an objective function makes it difficult to compare different models. As presented in [71], given a sufficient number of training samples and iterations, G learns to generate samples that resemble samples deriving from p_x. To ensure a sufficient quality of the generated samples, we require that they have the equivalent properties as the samples from p_x. We, therefore, build the following test scenarios.

We build a classifier that we choose to be architecturally a duplicate of that contained in the proposed MCGANs, see the discriminator and supervised loss function in figure 6.4. We then use an imbalanced training set, train the classifier on it, and validate the model on a balanced independent test set, see figure 6.6. This scenario we denote as S_{train}.

If G has properly learned to mimic p_x, the imbalanced training set, we can use the generated samples to oversample (balance) the training set, such that each class has the same number of samples. We then train a second classifier on the balanced set. We oversample before entering the training routine of the classifier and evaluate the model on the same test set as in S_{train}. We refer to this scenario as S_{MCGANs}. If p_G is similar to p_x, we expect no significant deviation in performance between S_{train} and S_{MCGANs}. We repeat each scenario ten times and report the errors in table 6.2. Interestingly,

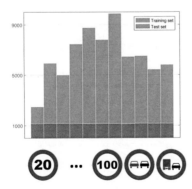

Figure 6.6: Dataset used for assessing quality of samples. Training and test set, each consisting of 11 classes: nine speed limit signs 20, 30, ..., 100 [km/h], and two no overtaking signs. Pictograms of those signs can be seen at the bottom of the histogram.

Scenario	Min. error	Mean error	Std.
S_{train}	4.7	4.8	0.1
S_{MCGANs}	4.3	4.8	0.2

Table 6.2: Assessment of the sample quality. Each scenario is repeated ten times and the minimum, mean, and standard division (Std.) of the achieved errors are reported.

we achieve in S_{MCGANs}, a reduced minimum error compared to S_{train}. This could be related to z_2, which affects the illumination and scale of the generated samples, as discussed in section 6.3.1. This is promising and we can expect further improvements by analyzing the corresponding components of z_2. Furthermore, the mean errors in both S_{train} and S_{MCGANs} are equal, although there is a greater standard deviation in the case of S_{MCGANs}; this highlights the need for a more careful procedure when creating the noise vectors for such purposes.

We observe MCGANs struggling because of the imbalance in the training data in a sense that, within the training process, samples with no clear membership to any class are created. We rectify this by randomly reducing the number of samples before each training epoch, such that all classes have exactly the same number of real samples during the entire GANs training.

6.3.3　Pixel to subpixel generator

MCGANs assign individual kernels to each channel in both models. These kernels learn independently from each other through shared abstract information. To explore if MCGANs allow transferring visual properties between channels, we discuss the use of the gray channel as input to the generator.

Using an input image for G, rather than the noise vector, is a considerably new technique and offers an entirely new spectrum of possible applications for GANs. Sixt et al. in [169], for example, use such a technique to generate a realistic bee marker from images from a 3D model. For this, the generator uses several differentiable functions that affect the blurriness, lighting, and other details of the input image. Isola et al. in [83] use such GANs for image-to-image translation. The authors use the architecture that is proposed in [148] to avoid a bottleneck structure that occurs when the generator assumes the shape of an encoder-decoder network. Our goal is to pass the gray channel to G and create the corresponding red channel. This could be used, for example, to increase the diversity within a given training set.

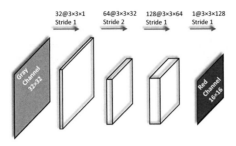

Figure 6.7: Architecture of the proposed pixel to subpixel generator. Blue arrows represent the convolution operations. Each operation is described by $N@H \times W \times C$, where N, H, W, C are number of neurons, height, width, and depth of the kernels, respectively.

Hence, we create a new architecture for G that inherently avoids creating a bottleneck structure.

We build G using convolutions and nonlinearities. To prevent any deterministic behavior, we provide noise to G using a dropout layer. As before, we further use batch normalization to accelerate the training process. Our new G is displayed in figure 6.7 and replaces the one from figure 6.4, while D remains architecturally unchanged. For passing its gradients to train G, D accepts as an input the real gray channel (input to G) and the real or generated red channel. The gradients derive from the binary loss function, see figure 6.4. Other loss functions do not contribute to the training process. The results of our trained model can be viewed in figure 6.8.

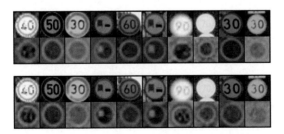

Figure 6.8: Top: real gray channel images with their corresponding red channels directly at their bottom. Bottom: real gray channel images with their corresponding generated red channels at their bottom.

As discussed in section 6.1.1.1, there is abundant successful research indicating that the use of synthetic data helps to improve the performance of classifiers and detectors. Chigorin et al. in [24], for example, confirm that the classifier performance can be significantly improved using real and synthetic data combined. For synthetic data, they use underlying pictograms of traffic signs and make them appear more naturally by employing effects and transformations such as rotation, scaling, shifting, applying salt and pepper noise.

This can also be achieved by modifying the generator from figure 6.7 such that it outputs both channels (gray and red) while being input with pictograms. The framework can learn to transfer visual properties, i.e., between real traffic signs and pictograms. Such a framework can also be used to make colored samples, captured from publicly available datasets, resemble samples captured with a different camera sensor.

6.4 Epilogue

We presented a general solution to how GANs can be adopted into domains where the input images have inherently different channels with different textures and resolutions. This was achieved, without loss of generality, by applying the concept of weight sharing on layers with a higher abstraction grade and separating kernels that are responsible for low-level details in both models (the discriminator and the generator). In this manner, both models can train kernels that are specialized in extracting and generating particular attributes from each channel without the requirement for exhaustive hyperparameter tuning. The multiple and separated channel representations have access to the same seed noise and loss functions. This allows the pipeline to be trained as a regular adversarial framework. We further demonstrated that the proposed solution can easily address the concept of conditions and confirmed via comprehensive experiments the effect of discrete and continuous conditions on the generated synthetic traffic signs. As an additional contribution, works that have successfully employed synthetic data in the field of traffic sign recognition have been discussed. Based on these and the presented study, we have suggested possible usage scenarios for the proposed framework and its generated synthetic data.

Chapter 7

Conclusion and Discussion

This thesis contributed to the research on image understanding for advanced driver-assistance systems (ADAS) by presenting novel approaches, designed for localizing pedestrians under unconstrained real-world conditions on low-consumption hardware. In addition, a novel approach was proposed to allow adversarial modeling frameworks to process image data generally used in ADAS. In this chapter, we conclude our study by recapitulating and discussing the problems, insights, and our contributions.

In chapter 2, we gave a systematic overview of today's detection frameworks and outlined the most prevalent approaches in machine learning and computer vision.

In chapter 3, we discussed relevant features that the image data must have in order to be suitable for pedestrian detection under unconstrained real-world conditions. We discussed two large and challenging datasets that are used to rank most state-of-the-art pedestrian detection methods and reviewed their evaluation metrics and methodologies. Both datasets were used in chapters 4 and 5 to compare our approaches to existing methods. We also introduced the red-clear-clear-clear traffic sign samples that we use in chapter 6 in our adversarial modeling framework.

Object detection has significantly improved in recent years by incorporating deep learning. In chapter 4, we investigated the utilization of deep learning in a cascaded boosted forest (CBF) pedestrian detector and aimed to economize on the computational cost of the arrangement. We introduced a novel detection pipeline that couples a CBF detector with a convolutional neural network (CNN). We demonstrated that the arrangement — that constitutes a two-stage cascade in a unified manner — runs virtually at the same frame rate as its underlying CBF detector while being significantly more accurate. The first constituent model, the CBF detector, produced a number of proposals that are highly likely to be pedestrians. The second constituent model, the CNN classifier, evaluated these samples. To accelerate the evaluation, we shared the feature pyramid that is constructed by the CBF detector and forwarded features that belong to the promising locations. We designed a small-sized CNN that could process the small proposal dimension and trained it from scratch. A crucial insight for achieving a sufficiently deep architecture, while keeping the number of trainable parameters low, was to omit pooling layers, that have historically been seen as an essential part of CNN architectures. We compensated for the missing pooling layers by using an intense training routine, an appropriate operational point, and an optimal location of the CNN classifier in the pipeline. In such arrangements, the second classifier benefits from a relatively high recall. In order to achieve an appropriate operational point in terms of runtime and recall, we increased the cascade threshold of our CBF detector and compensated the reduced recall by applying the CNN before the non-maximum suppression (NMS) algorithm. This allowed us to run the CBF detector at higher frames per second (FPS) while achieving a high recall, however, at the cost of a reduced precision. In terms of accuracy, our experiments showed that the CNN benefits from such a setup. We further introduced a novel strategy to discard proposals based on their height and distance from the camera which resulted in accelerating the runtime of the arrangement. Our approach ran on a single CPU core in real time with 30 FPS. In a variation of our approach, we demonstrated that CNNs can induce discriminative features from low-level channel features. Concretely, we gathered the CNN channels from various layers of our network and used them combined with the features generated from the CBF detector as input to a new CBF classifier. This resulted in an improvement at virtually no additional computational cost. Lastly, we presented a three-stage cascade which is built by extending our pipeline by a final component which was a deep pretrained CNN applied after NMS to keep its cost low. This three-stage cascade ranked as the fourth best-performing method reported on one of the challenging pedestrian datasets that are available online. The performance of a cascade tightly correlates with that of its first component as concluded in previous studies that we discussed in chapter 4 and demonstrated through our evaluations. Consequently, any improvement of the proposal generator is likely to improve the performance of our framework. Improvements may also be obtained by incorporating up-scaling of the template size, better handling of ignored regions, stronger features, improved weak classifiers, or boosting algorithm. Our small CNN, the second component of our cascade, mimics architecturally one of the smallest CNNs used for pedestrian detection and would generally improve with further evolution of neural networks. Using an enhanced architecture, optimizer, and initialization could possibly lead to an improved performance. Moreover, we only used the data available in the corresponding datasets, as stated in chapter 4. An additional training data commonly helps CNNs improving performance.

Machine learning models learn from implicit regularities existing in the training data, therefore, it is fundamental to choose suitable training samples, in terms of quality and quantity, to achieve a good result and prevent overfitting or underfitting. In chapter 5, we attempted to address potential problems that occur when training a CBF detector. Our main focus was to exploit the asymmetry in the dataset. The asymmetry existing in multiscale object detection datasets arises from the fact that the number of positive samples is limited whereas every location in an image that does not contain the object of interest can theoretically serve as a negative sample. We demonstrated that under the sliding window paradigm, which is predominantly used in CBF frameworks, each image provides several thousand negative samples, while images contain on average only a few positive samples. In practice, however, only a fraction of the available data is used, with the number of the negative training samples being around the same as that of the positive samples. In order to make use of the asymmetry in the dataset, we discussed the underlying boosting algorithm, usually a variant of AdaBoost, and concluded that the algorithm not only must be able to handle a highly imbalanced training set but must also benefit from a relatively high number of negative samples. A larger number of the negative samples, makes positive samples become more important to the algorithm. We identified that in practice the quality of the samples, especially that of the positive samples, and the training procedure prevent the classifier from benefiting from a higher number of negative samples. We proposed optimizations, in various aspects, based on the identified shortcomings of the training routine and the requirements of today's pedestrian detection benchmarks and introduced a routine that collects a suitable number of high-quality negative and positive samples. Our approach does not require data augmentation and makes, therefore, a grid search for finding the correct parameters of such techniques obsolete. Moreover, we validate that the unilaterally enlarged number of samples effectively prevents overfitting and allows increasing the model capacity without incurring the risk of performance reduction or poor generalization. While most prominent studies investigate developing more discriminative features, our research provides an orthogonal approach for training an accurate CBF classifiers in a detection framework. Therefore, our findings can be employed in existing CBF detection methods without decelerating the detector. We demonstrated that using our approach allows ranking as the second-best reported CBF detector on two challenging pedestrian datasets while using a relatively small number of simple aggregated channel features, and therefore, being multiple times faster than competitors. Analyzing the output of our detector revealed that major failures are double detections, the classification of vertical structures as pedestrians, and the classification of small pedestrians as backgrounds. This suggests that future works must consider these scenarios and integrate the insights stronger in the training routine. For instance, the classifier could be stronger penalized when making double detections. This would integrate the localization problem directly into the classifier training. Also, improvements of the non-maximum suppression algorithm or alternatives of that algorithm could help to improve the localization accuracy. In addition, enlarging the receptive field of the classifier might be a solution for preventing the confusion of vertical structures. The detection of small pedestrians could possibly be best improved by involving enhanced features. As discussed in chapter 4, an imbalanced training set can worsen the CNN performance. This could be related to the online training. Therefore, more investigations are needed on how to translate our findings into CNN training.

Deep models have shown impressive progression in recent years and currently lead most classification and detection benchmarks. The majority of practical machine learning approaches use supervised learning and since deep models have large capacities, saturating them sufficiently requires a large number of well-balanced labeled training samples. In many fields in ADAS, gathering labeled training data is costly and requires human effort. This is especially true in traffic sign recognition, because countries use different types of traffic signs and moreover, different traffic signs do not occur equally often. These difficulties suggest investigating the use of synthetic data or the involvement of advanced learning approaches. In chapter 6, we investigated the use of a promising approach, generative adversarial networks (GANs). Our primary contribution was the adaptation of the ordinary adversarial modeling framework to the image data provided in ADAS, the so-called red-clear-clear-clear images. We demonstrated that through architectural and algorithmic changes, GANs can process multiple channels that have different resolutions and textures, and generate real-looking red-clear-clear-clear traffic sign samples. Our solution is universal and can easily be applied to known adversarial modeling frameworks. In order to demonstrate that, we integrated known approaches into our framework that enabled the generator to create specific samples and even change the incisive attributes of the samples. Our secondary contribution was the study of relevant researches that successfully employ synthetic data in classification tasks. Based on the studies and detailed analyses and evaluations of our framework, we discussed future research directions. We concluded that GANs can contribute in multiple ways to the training of traffic sign classifiers. We demonstrated that a variation of our pipeline can be used to transfer visual properties. This pipeline, for instance, can be used to make pictograms or RGB traffic sign samples, taken from publicly available datasets, look like red-clear-clear-clear traffic sign samples. Our framework may also be used to generate hard negatives which will be template-like samples with natural-looking surroundings. Complementing the training data with these samples can encourage classifiers to focus only on information necessary for predicting the class label and as a result become insensitive to unimportant details. Finally, short-circuiting a trained generator with a trained classifier can allow identifying the vulnerabilities of the classifier in an unsupervised manner.

Bibliography

[1] Alexe, B., Deselaers, T., and Ferrari, V. (2012). Measuring the objectness of image windows. *IEEE transactions on pattern analysis and machine intelligence*, 34(11):2189–2202.

[2] Angelova, A., Krizhevsky, A., Vanhoucke, V., Ogale, A. S., and Ferguson, D. (2015). Real-time pedestrian detection with deep network cascades. In *BMVC*, pages 32–1.

[3] Appel, R., Fuchs, T., Dollár, P., and Perona, P. (2013). Quickly boosting decision trees – pruning underachieving features early. In *Proceedings of the 30th International Conference on Machine Learning (ICML-13)*, pages 594–602.

[4] Arjovsky, M. and Bottou, L. (2017). Towards principled methods for training generative adversarial networks. *arXiv preprint arXiv:1701.04862*.

[5] Arjovsky, M., Chintala, S., and Bottou, L. (2017). Wasserstein gan. *arXiv preprint arXiv:1701.07875*.

[6] Baumann, F., Ehlers, A., Vogt, K., and Rosenhahn, B. (2013). Cascaded random forest for fast object detection. In *Scandinavian Conference on Image Analysis*, pages 131–142. Springer.

[7] Belongie, S., Malik, J., and Puzicha, J. (2001). Matching shapes. In *Computer Vision, 2001. ICCV 2001. Proceedings. Eighth IEEE International Conference on*, volume 1, pages 454–461. IEEE.

[8] Benenson, R., Mathias, M., Timofte, R., and Van Gool, L. (2012a). Pedestrian detection at 100 frames per second. In *Computer Vision and Pattern Recognition (CVPR), 2012 IEEE Conference on*, pages 2903–2910. IEEE.

[9] Benenson, R., Mathias, M., Timofte, R., and Van Gool, L. (2012b). Pedestrian detection at 100 frames per second. http://aiweb.techfak.uni-bielefeld.de/content/bworld-robot-control-software/.

[10] Benenson, R., Mathias, M., Tuytelaars, T., and Van Gool, L. (2013). Seeking the strongest rigid detector. In *Computer Vision and Pattern Recognition (CVPR), 2013 IEEE Conference on*, pages 3666–3673. IEEE.

[11] Benenson, R., Omran, M., Hosang, J., and Schiele, B. (2014). Ten years of pedestrian detection, what have we learned? *arXiv preprint arXiv:1411.4304*.

[12] Bodla, N., Singh, B., Chellappa, R., and Davis, L. S. (2017). Soft-nms – improving object detection with one line of code. In *2017 IEEE International Conference on Computer Vision (ICCV)*, pages 5562–5570. IEEE.

[13] Bourdev, L. and Brandt, J. (2005). Robust object detection via soft cascade. In *Computer Vision and Pattern Recognition, 2005. CVPR 2005. IEEE Computer Society Conference on*, volume 2, pages 236–243. IEEE.

[14] Breiman, L. (1996). Bagging predictors. *Machine learning*, 24(2):123–140.

[15] Breiman, L. (2001). Random forests. *Machine learning*, 45(1):5–32.

[16] Breiman, L. et al. (1996). Heuristics of instability and stabilization in model selection. *The annals of statistics*, 24(6):2350–2383.

[17] Breiman, L., Friedman, J., Stone, C. J., and Olshen, R. A. (1984). *Classification and regression trees*. CRC press.

[18] Cai, Z., Fan, Q., Feris, R. S., and Vasconcelos, N. (2016). A unified multi-scale deep convolutional neural network for fast object detection. In *European Conference on Computer Vision*, pages 354–370. Springer.

[19] Cai, Z., Saberian, M., and Vasconcelos, N. (2015). Learning complexity-aware cascades for deep pedestrian detection. In *Proceedings of the IEEE International Conference on Computer Vision*, pages 3361–3369.

[20] Cao, J., Pang, Y., and Li, X. (2017). Learning multilayer channel features for pedestrian detection. *IEEE transactions on image processing*, 26(7):3210–3220.

[21] Chatfield, K., Simonyan, K., Vedaldi, A., and Zisserman, A. (2014). Return of the devil in the details: Delving deep into convolutional nets. *arXiv preprint arXiv:1405.3531*.

[22] Chen, X., Duan, Y., Houthooft, R., Schulman, J., Sutskever, I., and Abbeel, P. (2016). Infogan: Interpretable representation learning by information maximizing generative adversarial nets. In *Advances in Neural Information Processing Systems*, pages 2172–2180.

[23] Cheng, G. and Han, J. (2016). A survey on object detection in optical remote sensing images. *ISPRS Journal of Photogrammetry and Remote Sensing*, 117(117):11–28.

[24] Chigorin, A. and Konushin, A. (2013). A system for large-scale automatic traffic sign recognition and mapping. *CMRT13–City Models, Roads and Traffic*, 2013:13–17.

[25] Christopher, M. B. (2016). *Pattern Recognition and Machine Learning*. Springer-Verlag New York.

[26] Ciregan, D., Meier, U., and Schmidhuber, J. (2012). Multi-column deep neural networks for image classification. In *Computer Vision and Pattern Recognition (CVPR), 2012 IEEE Conference on*, pages 3642–3649. IEEE.

[27] Coates, A., Ng, A., and Lee, H. (2011). An analysis of single-layer networks in unsupervised feature learning. In *Proceedings of the fourteenth international conference on artificial intelligence and statistics*, pages 215–223.

[28] Costea, A. D., Vesa, A. V., and Nedevschi, S. (2015). Fast pedestrian detection for mobile devices. In *Intelligent Transportation Systems (ITSC), 2015 IEEE 18th International Conference on*, pages 2364–2369. IEEE.

[29] Dai, Z., Yang, Z., Yang, F., Cohen, W. W., and Salakhutdinov, R. (2017). Good semi-supervised learning that requires a bad gan. *arXiv preprint arXiv:1705.09783*.

[30] Dalal, N. and Triggs, B. (2005). Histograms of oriented gradients for human detection. In *Computer Vision and Pattern Recognition, 2005. CVPR 2005. IEEE Computer Society Conference on*, volume 1, pages 886–893. IEEE.

[31] Deng, J., Dong, W., Socher, R., Li, L.-J., Li, K., and Fei-Fei, L. (2009). ImageNet: A Large-Scale Hierarchical Image Database. In *CVPR09*.

[32] Denton, E. L., Chintala, S., Fergus, R., et al. (2015). Deep generative image models using a laplacian pyramid of adversarial networks. In *Advances in neural information processing systems*, pages 1486–1494.

[33] Dollár, P. Piotr's Computer Vision Matlab Toolbox (PMT). http://vision.ucsd.edu/~pdollar/toolbox/doc/index.html.

[34] Dollár, P., Appel, R., Belongie, S., and Perona, P. (2014). Fast feature pyramids for object detection. *IEEE Transactions on Pattern Analysis and Machine Intelligence*, 36(8):1532–1545.

[35] Dollár, P., Appel, R., and Kienzle, W. (2012a). Crosstalk cascades for frame-rate pedestrian detection. In *Computer Vision–ECCV 2012*, pages 645–659. Springer.

[36] Dollár, P., Belongie, S. J., and Perona, P. (2010). The fastest pedestrian detector in the west. *BMVC*, 2(3):7.

[37] Dollár, P., Tu, Z., Perona, P., and Belongie, S. (2009a). Integral channel features. *BMVC*.

[38] Dollár, P., Tu, Z., Tao, H., and Belongie, S. (2007). Feature mining for image classification. In *Computer Vision and Pattern Recognition, 2007. CVPR'07. IEEE Conference on*, pages 1–8. IEEE.

[39] Dollár, P., Wojek, C., Schiele, B., and Perona, P. (2009b). Pedestrian detection: A benchmark. In *Computer Vision and Pattern Recognition, 2009. CVPR 2009. IEEE Conference on*, pages 304–311. IEEE.

[40] Dollár, P., Wojek, C., Schiele, B., and Perona, P. (2012b). Pedestrian detection: An evaluation of the state of the art. *IEEE transactions on pattern analysis and machine intelligence*, 34(4):743–761.

[41] Dosovitskiy, A., Tobias Springenberg, J., and Brox, T. (2015). Learning to generate chairs with convolutional neural networks. In *Proceedings of the IEEE Conference on Computer Vision and Pattern Recognition*, pages 1538–1546.

[42] Dreyfus, S. (1962). The numerical solution of variational problems. *Journal of Mathematical Analysis and Applications*, 5(1):30–45.

[43] Efron, B. (1978). Regression and anova with zero-one data: Measures of residual variation. *Journal of the American Statistical Association*, 73(361):113–121.

[44] Efron, B. (1979). Bootstrap methods:{A} nother look at the jackknife. *Annals of Statistics*, 7:1–26.

[45] Enzweiler, M. and Gavrila, D. M. (2009). Monocular pedestrian detection: Survey and experiments. *IEEE transactions on pattern analysis and machine intelligence*, 31(12):2179–2195.

[46] Esposito, F., Malerba, D., Semeraro, G., and Kay, J. (1997). A comparative analysis of methods for pruning decision trees. *IEEE transactions on pattern analysis and machine intelligence*, 19(5):476–491.

[47] Ess, A., Leibe, B., Schindler, K., and Van Gool, L. (2008). A mobile vision system for robust multi-person tracking. In *Computer Vision and Pattern Recognition, 2008. CVPR 2008. IEEE Conference on*, pages 1–8. IEEE.

[48] Everingham, M., Van Gool, L., Williams, C. K., Winn, J., and Zisserman, A. (2010). The pascal visual object classes (voc) challenge. *International journal of computer vision*, 88(2):303–338.

[49] Felzenszwalb, P., McAllester, D., and Ramanan, D. (2008). A discriminatively trained, multiscale, deformable part model. In *Computer Vision and Pattern Recognition, 2008. CVPR 2008. IEEE Conference on*, pages 1–8. IEEE.

[50] Felzenszwalb, P. F., Girshick, R. B., McAllester, D., and Ramanan, D. (2010). Object detection with discriminatively trained part-based models. *IEEE transactions on pattern analysis and machine intelligence*, 32(9):1627–1645.

[51] Felzenszwalb, P. F. and Huttenlocher, D. P. (2004). Efficient graph-based image segmentation. *International journal of computer vision*, 59(2):167–181.

[52] Ferreira, A. J. and Figueiredo, M. A. (2012). Boosting algorithms: A review of methods, theory, and applications. In *Ensemble machine learning*, pages 35–85. Springer.

[53] Freeman, W. T. and Roth, M. (1995). Orientation histograms for hand gesture recognition. In *International workshop on automatic face and gesture recognition*, volume 12, pages 296–301.

[54] Freeman, W. T., Tanaka, K.-i., Ohta, J., and Kyuma, K. (1996). Computer vision for computer games. In *Automatic Face and Gesture Recognition, 1996., Proceedings of the Second International Conference on*, pages 100–105. IEEE.

[55] Freund, Y., Schapire, R., and Abe, N. (1999). A short introduction to boosting. *Journal-Japanese Society For Artificial Intelligence*, 14(771-780):1612.

[56] Freund, Y. and Schapire, R. E. (1995). A desicion-theoretic generalization of on-line learning and an application to boosting. In *European conference on computational learning theory*, pages 23–37. Springer.

[57] Freund, Y. and Schapire, R. E. (1997). A decision-theoretic generalization of on-line learning and an application to boosting. *Journal of computer and system sciences*, 55(1):119–139.

[58] Friedman, J., Hastie, T., and Tibshirani, R. (2001). *The elements of statistical learning*, volume 1. Springer series in statistics New York.

[59] Friedman, J., Hastie, T., Tibshirani, R., et al. (2000). Additive logistic regression: a statistical view of boosting (with discussion and a rejoinder by the authors). *The annals of statistics*, 28(2):337–407.

[60] Geiger, A., Lenz, P., and Urtasun, R. (2012). Are we ready for autonomous driving? the kitti vision benchmark suite. In *Computer Vision and Pattern Recognition (CVPR), 2012 IEEE Conference on*, pages 3354–3361. IEEE.

[61] Gerónimo, D., López, A., and Sappa, A. D. (2007). Computer vision approaches to pedestrian detection: visible spectrum survey. In *Iberian Conference on Pattern Recognition and Image Analysis*, pages 547–554. Springer.

[62] Ghorban, F., Marin, J., Su, Y., Colombo, A., and Kummert, A. (2018a). Aggregated channels network for real-time pedestrian detection. In *Tenth International Conference on Machine Vision (ICMV 2017)*, volume 10696.

[63] Ghorban, F., Milani, N., Schugk, D., Roese-Koerner, L., Su, Y., and Müller, D. (2017a). Methods of processing and generating image data in a connectionist network. European patent application no. 17208223.2.

[64] Ghorban, F., Milani, N., Schugk, D., Roese-Koerner, L., Su, Y., Müller, D., and Kummert, A. (2018b). Conditional multichannel generative adversarial networks with an application to traffic signs representation learning. *Progress in Artificial Intelligence*, pages 1–10.

[65] Ghorban, F., Su, Y., Meuter, M., and Kummert, A. (2017b). Insatiate boosted forest: Towards data exploitation in object detection. In *Intelligent Computer Communication and Processing (ICCP), 2017 13th IEEE International Conference on*, pages 331–338. IEEE.

[66] Ghorban, F., Su, Y., Tur, F. J. M., and Colombo, A. (2016). Image processing system to detect objects of interest. European patent application no. 16175330.6.

[67] Girshick, R. (2015). Fast r-cnn. In *Proceedings of the IEEE international conference on computer vision*, pages 1440–1448.

[68] Girshick, R., Donahue, J., Darrell, T., and Malik, J. (2014). Rich feature hierarchies for accurate object detection and semantic segmentation. In *Proceedings of the IEEE conference on computer vision and pattern recognition*, pages 580–587.

[69] Glorot, X., Bordes, A., and Bengio, Y. (2011). Deep sparse rectifier neural networks. In *Proceedings of the Fourteenth International Conference on Artificial Intelligence and Statistics*, pages 315–323.

[70] Goodfellow, I. (2016). Nips 2016 tutorial: Generative adversarial networks. *arXiv preprint arXiv:1701.00160*.

[71] Goodfellow, I., Pouget-Abadie, J., Mirza, M., Xu, B., Warde-Farley, D., Ozair, S., Courville, A., and Bengio, Y. (2014). Generative adversarial nets. In *Advances in neural information processing systems*, pages 2672–2680.

[72] Hagan, M. T., Demuth, H. B., Beale, M. H., et al. (1996). *Neural network design*, volume 20. Pws Pub. Boston.

[73] Haselhoff, A., Nunn, C., Müller, D., Meuter, M., and Roese-Koerner, L. (2017). Markov random field for image synthesis with an application to traffic sign recognition. In *Intelligent Vehicles Symposium (IV), 2017 IEEE*, pages 1407–1412. IEEE.

[74] He, K., Zhang, X., Ren, S., and Sun, J. (2016). Deep residual learning for image recognition. In *Proceedings of the IEEE conference on computer vision and pattern recognition*, pages 770–778.

[75] Ho, T. K. (1998). The random subspace method for constructing decision forests. *IEEE Transactions on Pattern Analysis and Machine Intelligence*, 20(8):832–844.

[76] Hosang, J., Benenson, R., Dollár, P., and Schiele, B. (2016a). What makes for effective detection proposals? *IEEE transactions on pattern analysis and machine intelligence*, 38(4):814–830.

[77] Hosang, J., Benenson, R., and Schiele, B. (2014). How good are detection proposals, really? In *25th British Machine Vision Conference*, pages 1–12. BMVA Press.

[78] Hosang, J., Benenson, R., and Schiele, B. (2016b). A convnet for non-maximum suppression. In *German Conference on Pattern Recognition*, pages 192–204. Springer.

[79] Hosang, J., Benenson, R., and Schiele, B. (2017). Learning non-maximum suppression. In *30th IEEE Conference on Computer Vision and Pattern Recognition*, pages 6469–6477. IEEE Computer Society.

[80] Hosang, J., Omran, M., Benenson, R., and Schiele, B. (2015). Taking a deeper look at pedestrians. In *Proceedings of the IEEE Conference on Computer Vision and Pattern Recognition*, pages 4073–4082.

[81] Huang, J., Rathod, V., Sun, C., Zhu, M., Korattikara, A., Fathi, A., Fischer, I., Wojna, Z., Song, Y., Guadarrama, S., et al. (2017). Speed/accuracy trade-offs for modern convolutional object detectors. In

Proceedings of the IEEE Conference on Computer Vision and Pattern Recognition, pages 7310–7311.

[82] Ioffe, S. and Szegedy, C. (2015). Batch normalization: Accelerating deep network training by reducing internal covariate shift. *arXiv preprint arXiv:1502.03167*.

[83] Isola, P., Zhu, J.-Y., Zhou, T., and Efros, A. A. (2016). Image-to-image translation with conditional adversarial networks. *arXiv preprint arXiv:1611.07004*.

[84] Janocha, K., Czarnecki, W. M., et al. (2017). On loss functions for deep neural networks in classification. *Schedae Informaticae*, 2016(Volume 25):4959.

[85] Kelley, H. J. (1960). Gradient theory of optimal flight paths. *Ars Journal*, 30(10):947–954.

[86] Khan, F. S., Anwer, R. M., Van De Weijer, J., Bagdanov, A. D., Vanrell, M., and Lopez, A. M. (2012). Color attributes for object detection. In *Computer Vision and Pattern Recognition (CVPR), 2012 IEEE Conference on*, pages 3306–3313. IEEE.

[87] Kim, T., Cha, M., Kim, H., Lee, J., and Kim, J. (2017). Learning to discover cross-domain relations with generative adversarial networks. *arXiv preprint arXiv:1703.05192*.

[88] Kingma, D. and Ba, J. (2014). Adam: A method for stochastic optimization. *arXiv preprint arXiv:1412.6980*.

[89] Kingma, D. P. and Welling, M. (2013). Auto-encoding variational bayes. *arXiv preprint arXiv:1312.6114*.

[90] Kleinberg, E. (1990). Stochastic discrimination. *Annals of Mathematics and Artificial intelligence*, 1(1):207–239.

[91] König, D. (2017). Deep learning for person detection in multi-spectral videos.

[92] Kotsiantis, S. B., Zaharakis, I., and Pintelas, P. (2007). Supervised machine learning: A review of classification techniques. *Emerging artificial intelligence applications in computer engineering*, 160:3–24.

[93] Krizhevsky, A. and Hinton, G. (2009). Learning multiple layers of features from tiny images. *Technical report, University of Toronto*.

[94] Krizhevsky, A., Sutskever, I., and Hinton, G. E. (2012). Imagenet classification with deep convolutional neural networks. In *Advances in neural information processing systems*, pages 1097–1105.

[95] Krogh, A. and Hertz, J. A. (1992). A simple weight decay can improve generalization. In *Advances in neural information processing systems*, pages 950–957.

[96] Kumar, A., Sattigeri, P., and Fletcher, T. (2017). Semi-supervised learning with gans: Manifold invariance with improved inference. In *Advances in Neural Information Processing Systems*, pages 5540–5550.

[97] Kuncheva, L. I. (2004). *Combining pattern classifiers: methods and algorithms*. John Wiley & Sons.

[98] Kuncheva, L. I. and Whitaker, C. J. (2003). Measures of diversity in classifier ensembles and their relationship with the ensemble accuracy. *Machine learning*, 51(2):181–207.

[99] Lampert, C. H., Blaschko, M. B., and Hofmann, T. (2009). Efficient subwindow search: A branch and bound framework for object localization. *IEEE transactions on pattern analysis and machine intelligence*, 31(12):2129–2142.

[100] Laptev, I. (2006). Improvements of object detection using boosted histograms. In *BMVC*, volume 3, pages 949–958.

[101] LeCun, Y., Boser, B., Denker, J., Henderson, D., Howard, R., Hubbard, W., and Jackel, L. (1989). Backpropagation applied to handwritten zip code recognition. *Neural Computation*, 1:541–551.

[102] Ledig, C., Theis, L., Huszár, F., Caballero, J., Cunningham, A., Acosta, A., Aitken, A., Tejani, A., Totz, J., Wang, Z., et al. (2016). Photo-realistic single image super-resolution using a generative adversarial network. *arXiv preprint*.

[103] Lenc, K. and Vedaldi, A. (2015). R-cnn minus r. *arXiv preprint arXiv:1506.06981*.

[104] Li, J., Liang, X., Shen, S., Xu, T., Feng, J., and Yan, S. (2017). Scale-aware fast r-cnn for pedestrian detection. *IEEE Transactions on Multimedia*.

[105] Lienhart, R. and Maydt, J. (2002). An extended set of haar-like features for rapid object detection. In *Image Processing. 2002. Proceedings. 2002 International Conference on*, volume 1, pages I–I. IEEE.

[106] Lowe, D. G. (1999). Object recognition from local scale-invariant features. In *Computer vision, 1999. The proceedings of the seventh IEEE international conference on*, volume 2, pages 1150–1157. Ieee.

[107] Lowe, D. G. (2004). Distinctive image features from scale-invariant keypoints. *International journal of computer vision*, 60(2):91–110.

[108] Luo, H. (2005). Optimization design of cascaded classifiers. In *Computer Vision and Pattern Recognition, 2005. CVPR 2005. IEEE Computer Society Conference on*, volume 1, pages 480–485. IEEE.

[109] Luo, P., Tian, Y., Wang, X., and Tang, X. (2014). Switchable deep network for pedestrian detection. In *Proceedings of the IEEE Conference on Computer Vision and Pattern Recognition*, pages 899–906.

[110] Maas, A. L., Hannun, A. Y., and Ng, A. Y. (2013). Rectifier nonlinearities improve neural network acoustic models. In *Proc. icml*, volume 30, page 3.

[111] Mao, X., Li, Q., Xie, H., Lau, R. Y., Wang, Z., and Smolley, S. P. (2016). Least squares generative adversarial networks. *arXiv preprint ArXiv:1611.04076*.

[112] Marin, J., Vázquez, D., López, A. M., Amores, J., and Leibe, B. (2013). Random forests of local experts for pedestrian detection. In *Proceedings of the IEEE International Conference on Computer Vision*, pages 2592–2599.

[113] Masnadi-Shirazi, H. and Vasconcelos, N. (2007). Asymmetric boosting. In *Proceedings of the 24th international conference on Machine learning*, pages 609–619. ACM.

[114] Mathias, M., Timofte, R., Benenson, R., and Van Gool, L. (2013). Traffic sign recognition – how far are we from the solution? In *Neural Networks (IJCNN), The 2013 International Joint Conference on*, pages 1–8. IEEE.

[115] McCulloch, W. S. and Pitts, W. (1943). A logical calculus of the ideas immanent in nervous activity. *The bulletin of mathematical biophysics*, 5(4):115–133.

[116] Mingers, J. (1989). An empirical comparison of selection measures for decision-tree induction. *Machine learning*, 3(4):319–342.

[117] Minsky, M. and Papert, S. (1969). *Perceptrons: An Introduction to Computational Geometry*. MIT press.

[118] Mirza, M. and Osindero, S. (2014). Conditional generative adversarial nets. *arXiv preprint arXiv:1411.1784*.

[119] Mogelmose, A., Trivedi, M. M., and Moeslund, T. B. (2012). Learning to detect traffic signs: Comparative evaluation of synthetic and real-world datasets. In *Pattern Recognition (ICPR), 2012 21st International Conference on*, pages 3452–3455. IEEE.

[120] Moiseev, B., Konev, A., Chigorin, A., and Konushin, A. (2013). Evaluation of traffic sign recognition methods trained on synthetically generated data. In *International Conference on Advanced Concepts for Intelligent Vision Systems*, pages 576–583. Springer.

[121] Murthy, S. K. (1998). Automatic construction of decision trees from data: A multi-disciplinary survey. *Data mining and knowledge discovery*, 2(4):345–389.

[122] Nam, W., Dollár, P., and Han, J. H. (2014). Local decorrelation for improved pedestrian detection. In *Advances in Neural Information Processing Systems*, pages 424–432.

[123] Odena, A. (2016). Semi-supervised learning with generative adversarial networks. *arXiv preprint arXiv:1606.01583*.

[124] Odena, A., Olah, C., and Shlens, J. (2016). Conditional image synthesis with auxiliary classifier gans. *arXiv preprint arXiv:1610.09585*.

[125] Ohn-Bar, E. and Trivedi, M. M. (2016). To boost or not to boost? on the limits of boosted trees for object detection. In *Pattern Recognition (ICPR), 2016 23rd International Conference on*, pages 3350–3355. IEEE.

[126] Oord, A. v. d., Kalchbrenner, N., and Kavukcuoglu, K. (2016). Pixel recurrent neural networks. *arXiv preprint arXiv:1601.06759*.

[127] Ouyang, W. and Wang, X. (2012). A discriminative deep model for pedestrian detection with occlusion handling. In *Computer Vision and Pattern Recognition (CVPR), 2012 IEEE Conference on*, pages 3258–3265. IEEE.

[128] Ouyang, W. and Wang, X. (2013a). Joint deep learning for pedestrian detection. In *Computer Vision (ICCV), 2013 IEEE International Conference on*, pages 2056–2063. IEEE.

[129] Ouyang, W. and Wang, X. (2013b). Single-pedestrian detection aided by multi-pedestrian detection. In *Proceedings of the IEEE Conference on Computer Vision and Pattern Recognition*, pages 3198–3205.

[130] Ouyang, W., Zeng, X., and Wang, X. (2013). Modeling mutual visibility relationship in pedestrian detection. In *Proceedings of the IEEE Conference on Computer Vision and Pattern Recognition*, pages 3222–3229.

[131] Paganini, M., de Oliveira, L., and Nachman, B. (2018). Accelerating science with generative adversarial networks: An application to 3d particle showers in multilayer calorimeters. *Physical review letters*, 120(4):042003.

[132] Paisitkriangkrai, S., Shen, C., and van den Hengel, A. (2016). Pedestrian detection with spatially pooled features and structured ensemble learning. *IEEE transactions on pattern analysis and machine intelligence*, 38(6):1243–1257.

[133] Papageorgiou, C. P., Oren, M., and Poggio, T. (1998). A general framework for object detection. In *Computer vision, 1998. sixth international conference on*, pages 555–562. IEEE.

[134] Park, D., Ramanan, D., and Fowlkes, C. (2010). Multiresolution models for object detection. *Computer Vision–ECCV 2010*, pages 241–254.

[135] Park, K.-Y. and Hwang, S.-Y. (2014). Robust range estimation with a monocular camera for vision-based forward collision warning system. *The Scientific World Journal*, 2014.

[136] Perronnin, F. and Dance, C. (2007). Fisher kernels on visual vocabularies for image categorization. In *Computer Vision and Pattern Recognition, 2007. CVPR'07. IEEE Conference on*, pages 1–8. IEEE.

[137] Plaut, D. C., Nowlan, S. J., and Hinton, G. E. (1987). Experiments on learning by back propagation. Technical report, CARNEGIE-MELLON UNIV PITTSBURGH PA DEPT OF COMPUTER SCIENCE.

[138] Ponce, J., Berg, T., Everingham, M., Forsyth, D., Hebert, M., Lazebnik, S., Marszalek, M., Schmid, C., Russell, B., Torralba, A., et al. (2006). Dataset issues in object recognition. *Toward category-level object recognition*, pages 29–48.

[139] Porikli, F. (2005). Integral histogram: A fast way to extract histograms in cartesian spaces. In *Computer Vision and Pattern Recognition, 2005. CVPR 2005. IEEE Computer Society Conference on*, volume 1, pages 829–836. IEEE.

[140] Radford, A., Metz, L., and Chintala, S. (2015). Unsupervised representation learning with deep convolutional generative adversarial networks. *arXiv preprint arXiv:1511.06434*.

[141] Rahtu, E., Kannala, J., and Blaschko, M. (2011). Learning a category independent object detection cascade. In *Computer Vision (ICCV), 2011 IEEE International Conference on*, pages 1052–1059. IEEE.

[142] Redmon, J., Divvala, S., Girshick, R., and Farhadi, A. (2016). You only look once: Unified, real-time object detection. In *Proceedings of the IEEE conference on computer vision and pattern recognition*, pages 779–788.

[143] Reed, S., Akata, Z., Yan, X., Logeswaran, L., Schiele, B., and Lee, H. (2016). Generative adversarial text to image synthesis. *arXiv preprint arXiv:1605.05396*.

[144] Ren, S., He, K., Girshick, R., and Sun, J. (2015). Faster r-cnn: Towards real-time object detection with region proposal networks. In *Advances in neural information processing systems*, pages 91–99.

[145] Ribeiro, D., Mateus, A., Miraldo, P., and Nascimento, J. C. (2017a). A real-time deep learning pedestrian detector for robot navigation. In *Autonomous Robot Systems and Competitions (ICARSC), 2017 IEEE International Conference on*, pages 165–171. IEEE.

[146] Ribeiro, D., Nascimento, J. C., Bernardino, A., and Carneiro, G. (2017b). Improving the performance of pedestrian detectors using convolutional learning. *Pattern Recognition*, 61:641–649.

[147] Rojas, R. (2013). *Neural networks: a systematic introduction*. Springer Science & Business Media.

[148] Ronneberger, O., Fischer, P., and Brox, T. (2015). U-net: Convolutional networks for biomedical image segmentation. In *International Conference on Medical Image Computing and Computer-Assisted Intervention*, pages 234–241. Springer.

[149] Rosenblatt, F. (1958). The perceptron: A probabilistic model for information storage and organization in the brain. *Psychological review*, 65(6):386.

[150] Rosenblatt, F. (1962). *Principles of neurodynamics*. Spartan Book.

[151] Rowley, H. A., Baluja, S., and Kanade, T. (1998). Neural network-based face detection. *IEEE Transactions on pattern analysis and machine intelligence*, 20(1):23–38.

[152] Ruder, S. (2016). An overview of gradient descent optimization algorithms. *arXiv preprint arXiv:1609.04747*.

[153] Ruderman, D. L. (1994). The statistics of natural images. *Network: computation in neural systems*, 5(4):517–548.

[154] Rumelhart, D. E., Hinton, G. E., and Williams, R. J. (1985). Learning internal representations by error propagation. Technical report, California Univ San Diego La Jolla Inst for Cognitive Science.

[155] Rumelhart, D. E., Hinton, G. E., and Williams, R. J. (1986). Learning representations by back-propagating errors. *nature*, 323(6088):533.

[156] Saberian, M. J. and Vasconcelos, N. (2012). Learning optimal embedded cascades. *IEEE transactions on pattern analysis and machine intelligence*, 34(10):2005–2018.

[157] Saberian, M. J. and Vasconcelos, N. (2014). Boosting algorithms for detector cascade learning. *Journal of Machine Learning Research*, 15(1):2569–2605.

[158] Salimans, T., Goodfellow, I., Zaremba, W., Cheung, V., Radford, A., and Chen, X. (2016). Improved techniques for training gans. In *Advances in Neural Information Processing Systems*, pages 2234–2242.

[159] Schapire, R. E. (1990). The strength of weak learnability. *Machine learning*, 5(2):197–227.

[160] Schapire, R. E. (2013). Explaining adaboost. In *Empirical inference*, pages 37–52. Springer.

[161] Schapire, R. E. and Freund, Y. (2012). *Boosting: Foundations and algorithms*. MIT press.

[162] Schapire, R. E., Freund, Y., Bartlett, P., Lee, W. S., et al. (1998). Boosting the margin: A new explanation for the effectiveness of voting methods. *The annals of statistics*, 26(5):1651–1686.

[163] Schapire, R. E. and Singer, Y. (1999). Improved boosting algorithms using confidence-rated predictions. *Machine learning*, 37(3):297–336.

[164] Sermanet, P., Kavukcuoglu, K., Chintala, S., and LeCun, Y. (2013). Pedestrian detection with unsupervised multi-stage feature learning. In *Proceedings of the IEEE Conference on Computer Vision and Pattern Recognition*, pages 3626–3633.

[165] Serre, T., Heisele, B., Mukherjee, S., and Poggio, T. (2000). Feature selection for face detection.

[166] Shashua, A., Gdalyahu, Y., and Hayun, G. (2004). Pedestrian detection for driving assistance systems: Single-frame classification and system level performance. In *Intelligent Vehicles Symposium, 2004 IEEE*, pages 1–6. IEEE.

[167] Shechtman, E. and Irani, M. (2007). Matching local self-similarities across images and videos. In *Computer Vision and Pattern Recognition, 2007. CVPR'07. IEEE Conference on*, pages 1–8. IEEE.

[168] Simonyan, K. and Zisserman, A. (2014). Very deep convolutional networks for large-scale image recognition. *arXiv preprint arXiv:1409.1556*.

[169] Sixt, L. (2016). Rendergan: Generating realistic labeled data – with an application on decoding bee tags. *unpublished Bachelor Thesis, Freie Universität Berlin*.

[170] Springenberg, J. T. (2015). Unsupervised and semi-supervised learning with categorical generative adversarial networks. *arXiv preprint arXiv:1511.06390*.

[171] Springenberg, J. T., Dosovitskiy, A., Brox, T., and Riedmiller, M. (2014). Striving for simplicity: The all convolutional net. *arXiv preprint arXiv:1412.6806*.

[172] Srivastava, N., Hinton, G., Krizhevsky, A., Sutskever, I., and Salakhutdinov, R. (2014). Dropout: A simple way to prevent neural networks from overfitting. *The Journal of Machine Learning Research*, 15(1):1929–1958.

[173] Stallkamp, J., Schlipsing, M., Salmen, J., and Igel, C. (2012). Man vs. computer: Benchmarking machine learning algorithms for traffic sign recognition. *Neural networks*, 32:323–332.

[174] Sun, J., Rehg, J. M., and Bobick, A. (2004). Automatic cascade training with perturbation bias. In *Computer Vision and Pattern Recognition, 2004. CVPR 2004. Proceedings of the 2004 IEEE Computer Society Conference on*, volume 2, pages II–II. IEEE.

[175] Sun, Y., Wong, A. K., and Wang, Y. (2005). Parameter inference of cost-sensitive boosting algorithms. In *MLDM*, pages 21–30. Springer.

[176] Sung, K.-K. and Poggio, T. (1998). Example-based learning for view-based human face detection. *IEEE Transactions on pattern analysis and machine intelligence*, 20(1):39–51.

[177] Szegedy, C., Liu, W., Jia, Y., Sermanet, P., Reed, S., Anguelov, D., Erhan, D., Vanhoucke, V., and Rabinovich, A. (2015). Going deeper with convolutions. In *Proceedings of the IEEE conference on computer vision and pattern recognition*, pages 1–9.

[178] Tibshirani, R. (1996). *Bias, variance and prediction error for classification rules*. University of Toronto, Department of Statistics.

[179] Tomizuka, M., Tai, M., Wang, J.-Y., and Hingwe, P. (1999). Automated lane guidance of commercial vehicles. In *Control Applications, 1999. Proceedings of the 1999 IEEE International Conference on*, volume 2, pages 1359–1364. IEEE.

[180] Tu, Z. (2005). Probabilistic boosting-tree: Learning discriminative models for classification, recognition, and clustering. In *Computer Vision, 2005. ICCV 2005. Tenth IEEE International Conference on*, volume 2, pages 1589–1596. IEEE.

[181] Uijlings, J. R., Van De Sande, K. E., Gevers, T., and Smeulders, A. W. (2013). Selective search for object recognition. *International journal of computer vision*, 104(2):154–171.

[182] Urmson, C., Anhalt, J., Bagnell, D., Baker, C., Bittner, R., Clark, M., Dolan, J., Duggins, D., Galatali, T., Geyer, C., et al. (2008). Autonomous driving in urban environments: Boss and the urban challenge. *Journal of Field Robotics*, 25(8):425–466.

[183] Vapnik, V. (1963). Pattern recognition using generalized portrait method. *Automation and remote control*, 24:774–780.

[184] Vapnik, V. and Chervonenkis, A. (1964). A note on one class of perceptrons. *Automation and remote control*, 25(1):103.

[185] Varga, R., Vesa, A. V., Jeong, P., and Nedevschi, S. (2014). Real-time pedestrian detection in urban scenarios. In *Intelligent Computer Communication and Processing (ICCP), 2014 IEEE International Conference on*, pages 113–118. IEEE.

[186] Vedaldi, A. and Lenc, K. (2015a). Matconvnet – convolutional neural networks for matlab. In *Proceeding of the ACM Int. Conf. on Multimedia*.

[187] Vedaldi, A. and Lenc, K. (2015b). Matconvnet: Convolutional neural networks for matlab. In *Proceedings of the 23rd ACM international conference on Multimedia*, pages 689–692. ACM.

[188] Viola, P. and Jones, M. (2001). Rapid object detection using a boosted cascade of simple features. In *Computer Vision and Pattern Recognition, 2001. CVPR 2001. Proceedings of the 2001 IEEE Computer Society Conference on*, volume 1, pages I–I. IEEE.

[189] Viola, P. and Jones, M. (2002). Fast and robust classification using asymmetric adaboost and a detector cascade. In *Advances in neural information processing systems*, pages 1311–1318.

[190] Viola, P. and Jones, M. J. (2004). Robust real-time face detection. *International journal of computer vision*, 57(2):137–154.

[191] Walk, S., Majer, N., Schindler, K., and Schiele, B. (2010). New features and insights for pedestrian detection. In *Computer vision and pattern recognition (CVPR), 2010 IEEE conference on*, pages 1030–1037. IEEE.

[192] Wang, X., Han, T. X., and Yan, S. (2009). An hog-lbp human detector with partial occlusion handling. In *Computer Vision, 2009 IEEE 12th International Conference on*, pages 32–39. IEEE.

[193] Wang, X., Shrivastava, A., and Gupta, A. (2017). A-fast-rcnn: Hard positive generation via adversary for object detection. *arXiv preprint arXiv:1704.03414*.

[194] Wang, X., Yang, M., Zhu, S., and Lin, Y. (2013). Regionlets for generic object detection. In *Proceedings of the IEEE International Conference on Computer Vision*, pages 17–24.

[195] Werbos, P. J. (1974). Beyond regression: New tools for prediction and analysis in the behavioral sciences. *Doctoral Dissertation, Applied Mathematics, Harvard University, MA*.

[196] Wojek, C., Walk, S., and Schiele, B. (2009). Multi-cue onboard pedestrian detection. In *Computer Vision and Pattern Recognition, 2009. CVPR 2009. IEEE Conference on*, pages 794–801. IEEE.

[197] Wu, B. and Nevatia, R. (2005). Detection of multiple, partially occluded humans in a single image by bayesian combination of edgelet part detectors. In *Computer Vision, 2005. ICCV 2005. Tenth IEEE International Conference on*, volume 1, pages 90–97. IEEE.

[198] Xiang, Y., Choi, W., Lin, Y., and Savarese, S. (2017). Subcategory-aware convolutional neural networks for object proposals and detection. In *Applications of Computer Vision (WACV), 2017 IEEE Winter Conference on*, pages 924–933. IEEE.

[199] Xiao, R., Zhu, H., Sun, H., and Tang, X. (2007). Dynamic cascades for face detection. In *Computer Vision, 2007. ICCV 2007. IEEE 11th International Conference on*, pages 1–8. IEEE.

[200] Xiao, R., Zhu, L., and Zhang, H.-J. (2003). Boosting chain learning for object detection. In *Computer Vision, 2003. Proceedings. Ninth IEEE International Conference on*, pages 709–715. IEEE.

[201] Yan, J., Zhang, X., Lei, Z., Liao, S., and Li, S. Z. (2013). Robust multi-resolution pedestrian detection in traffic scenes. In *Proceedings of the IEEE Conference on Computer Vision and Pattern Recognition*, pages 3033–3040.

[202] Yang, B., Yan, J., Lei, Z., and Li, S. Z. (2015). Convolutional channel features. In *Proceedings of the IEEE international conference on computer vision*, pages 82–90.

[203] Yang, F., Choi, W., and Lin, Y. (2016). Exploit all the layers: Fast and accurate cnn object detector with scale dependent pooling and cascaded rejection classifiers. In *Proceedings of the IEEE Conference on Computer Vision and Pattern Recognition*, pages 2129–2137.

[204] Yoo, D., Park, S., Lee, J.-Y., Paek, A. S., and So Kweon, I. (2015). Attentionnet: Aggregating weak directions for accurate object detection. In *Proceedings of the IEEE International Conference on Computer Vision*, pages 2659–2667.

[205] Zeng, X., Ouyang, W., and Wang, X. (2013). Multi-stage contextual deep learning for pedestrian detection. In *Proceedings of the IEEE International Conference on Computer Vision*, pages 121–128.

[206] Zhang, C. and Viola, P. A. (2008). Multiple-instance pruning for learning efficient cascade detectors. In *Advances in neural information processing systems*, pages 1681–1688.

[207] Zhang, L., Lin, L., Liang, X., and He, K. (2016a). Is faster r-cnn doing well for pedestrian detection? In *European Conference on Computer Vision*, pages 443–457. Springer.

[208] Zhang, S., Bauckhage, C., and Cremers, A. B. (2014). Informed haarlike features improve pedestrian detection. In *Proceedings of the IEEE conference on computer vision and pattern recognition*, pages 947–954.

[209] Zhang, S., Benenson, R., Omran, M., Hosang, J., and Schiele, B. (2016b). How far are we from solving pedestrian detection? In *Proceedings of the IEEE Conference on Computer Vision and Pattern Recognition*, pages 1259–1267.

[210] Zhang, S., Benenson, R., and Schiele, B. (2015). Filtered channel features for pedestrian detection. In *Computer Vision and Pattern Recognition (CVPR), 2015 IEEE Conference on*, pages 1751–1760. IEEE.

[211] Zhang, W., Sun, J., and Tang, X. (2008). Cat head detection-how to effectively exploit shape and texture features. *Computer Vision–ECCV 2008*, pages 802–816.

[212] Zheng, Z., Zheng, L., and Yang, Y. (2017). Unlabeled samples generated by gan improve the person re-identification baseline in vitro. *arXiv preprint arXiv:1701.07717*.

[213] Zhou, Z.-H. (2012). *Ensemble methods: foundations and algorithms.* CRC press.

[214] Zhu, Q., Yeh, M.-C., Cheng, K.-T., and Avidan, S. (2006). Fast human detection using a cascade of histograms of oriented gradients. In *Computer Vision and Pattern Recognition, 2006 IEEE Computer Society Conference on,* volume 2, pages 1491–1498. IEEE.

[215] Zitnick, C. L. and Dollár, P. (2014). Edge boxes: Locating object proposals from edges. In *European Conference on Computer Vision,* pages 391–405. Springer.

In der Reihe *Studien zur Mustererkennung,*
herausgegeben von
Prof. Dr.-Ing Heinricht Niemann und Herrn Prof. Dr.-Ing. Elmar Nöth
sind bisher erschienen:

1	Jürgen Haas	Probabilistic Methods in Linguistic Analysis
		ISBN 978-3-89722-565-7, 2000, 260 S. 40.50 €
2	Manuela Boros	Partielles robustes Parsing spontansprachlicher Dialoge am Beispiel von Zugauskunftdialogen
		ISBN 978-3-89722-600-5, 2001, 264 S. 40.50 €
3	Stefan Harbeck	Automatische Verfahren zur Sprachdetektion, Landessprachenerkennung und Themendetektion
		ISBN 978-3-89722-766-8, 2001, 260 S. 40.50 €
4	Julia Fischer	Ein echtzeitfähiges Dialogsystem mit iterativer Ergebnisoptimierung
		ISBN 978-3-89722-867-2, 2002, 222 S. 40.50 €
5	Ulrike Ahlrichs	Wissensbasierte Szenenexploration auf der Basis erlernter Analysestrategien
		ISBN 978-3-89722-904-4, 2002, 165 S. 40.50 €
6	Florian Gallwitz	Integrated Stochastic Models for Spontaneous Speech Recognition
		ISBN 978-3-89722-907-5, 2002, 196 S. 40.50 €
7	Uwe Ohler	Computational Promoter Recognition in Eukaryotic Genomic DNA
		ISBN 978-3-89722-988-4, 2002, 206 S. 40.50 €
8	Richard Huber	Prosodisch-linguistische Klassifikation von Emotion
		ISBN 978-3-89722-984-6, 2002, 293 S. 40.50 €

9	Volker Warnke	Integrierte Segmentierung und Klassifikation von Äußerungen und Dialogakten mit heterogenen Wissensquellen
		ISBN 978-3-8325-0254-6, 2003, 182 S. 40.50 €
10	Michael Reinhold	Robuste, probabilistische, erscheinungsbasierte Objekterkennung
		ISBN 978-3-8325-0476-2, 2004, 283 S. 40.50 €
11	Matthias Zobel	Optimale Brennweitenwahl für die multiokulare Objektverfolgung
		ISBN 978-3-8325-0496-0, 2004, 292 S. 40.50 €
12	Bernd Ludwig	Ein konfigurierbares Dialogsystem für Mensch-Maschine-Interaktion in gesprochener Sprache
		ISBN 978-3-8325-0497-7, 2004, 230 S. 40.50 €
13	Rainer Deventer	Modeling and Control of Static and Dynamic Systems with Bayesian Networks
		ISBN 978-3-8325-0521-9, 2004, 195 S. 40.50 €
14	Jan Buckow	Multilingual Prosody in Automatic Speech Understanding
		ISBN 978-3-8325-0581-3, 2004, 164 S. 40.50 €
15	Klaus Donath	Automatische Segmentierung und Analyse von Blutgefäßen
		ISBN 978-3-8325-0642-1, 2004, 210 S. 40.50 €
16	Axel Walthelm	Sensorbasierte Lokalisations-Algorithmen für mobile Service-Roboter
		ISBN 978-3-8325-0691-9, 2004, 200 S. 40.50 €
17	Ricarda Dormeyer	Syntaxanalyse auf der Basis der Dependenzgrammatik
		ISBN 978-3-8325-0723-7, 2004, 200 S. 40.50 €
18	Michael Levit	Spoken Language Understanding without Transcriptions in a Call Center Scenario
		ISBN 978-3-8325-0930-9, 2005, 249 S. 40.50 €

39	Chen Li	Content-based Microscopic Image Analysis

ISBN 978-3-8325-4253-5, 2016, 196 S.   36.50 €

| 40 | Christian Feinen | Object Representation and Matching Based on Skeletons and Curves |

ISBN 978-3-8325-4257-3, 2016, 260 S.   50.50 €

| 41 | Juan Rafael Orozco-Arroyave | Analysis of Speech of People with Parkinson's Disease |

ISBN 978-3-8325-4361-7, 2016, 138 S.   38.00 €

| 42 | Cong Yang | Object Shape Generation, Representation and Matching |

ISBN 978-3-8325-4399-0, 2016, 194 S.   36.50 €

| 43 | Florian Hönig | Automatic Assessment of Prosody in Second Language Learning |

ISBN 978-3-8325-4567-3, 2017, 256 S.   38.50 €

| 44 | Zeyd Boukhers | 3D Trajectory Extraction from 2D Videos for Human Activity Analysis |

ISBN 978-3-8325-4583-3, 2017, 152 S.   35.00 €

| 45 | Muhammad Hassan Khan | Human Activity Analysis in Visual Surveillance and Healthcare |

ISBN 978-3-8325-4807-0, 2018, 156 S.   35.00 €

| 46 | Lukas Köping | Probabilistic Fusion of Multiple Distributed Sensors |

ISBN 978-3-8325-4827-8, 2018, 170 S.   46.50 €

| 47 | Farzin Ghorban | Machine Learning in Advanced Driver-Assistance Systems - Contributions to Pedestrian Detection and Adversarial Modeling |

ISBN 978-3-8325-4874-2, 2019, 153 S.   40.50 €

Alle erschienenen Bücher können unter der angegebenen ISBN im Buchhandel oder direkt beim Logos Verlag Berlin (www.logos-verlag.de, Fax: 030 - 42 85 10 92) bestellt werden.